Technological Concepts and Mathematical Models in the Evolution of Modern Engineering Systems

Controlling · Managing · Organizing

Edited by
Mario Lucertini
Ana Millán Gasca
Fernando Nicolò

Springer Basel AG

Editors:

Dr. Ana Millán Gasca
Dipartimento di Matematica
Università di Roma "La Sapienza"
Piazzale Aldo Moro 2
I-00185 Roma
Email: millan@mat.uniroma1.it

Prof. Fernando Nicolò
Dipartimento di Informatica e Automazione
Università di Roma 3
Via della Vasca Navale 79
I-00146 Roma
Email: nicolo@uniroma3.it

Prof. Mario Lucertini was a member of
Dipartimento di Informatica, Sistemi e Produzione
Università di Roma "Tor Vergata"

Published with the support of the UNESCO Venice Office – Regional Office for Science & Technology in Europe

A CIP catalogue record for this book is available from the Library of Congress, Washington D.C., USA

Deutsche Bibliothek Cataloging-in-Publication Data
Die Deutsche Bibliothek lists this publication in the Deutsche Nationalbibliografie;
detailed bibliographic data is available in the Internet at http://dnb.ddb.de.

ISBN 978-3-0348-9633-7   ISBN 978-3-0348-7951-4 (eBook)
DOI 10.1007/978-3-0348-7951-4
© 2004 Springer Basel AG
Originally published by Birkhäuser Verlag in 2004
Softcover reprint of the hardcover 1st edition 2004
Printed on acid-free paper produced from chlorine-free pulp. TCF ∞
Cover design: Micha Lotrovsky, CH-4106 Therwil, Switzerland
Cover illustration: Fernand Léger: *La Ville* (1919) [© 2003 ProLitteris, Zürich]

ISBN 978-3-0348-9633-7

9 8 7 6 5 4 3 2 1   www.birkhäuser-science.com

Table of Contents

Introduction

M. Lucertini, A. Millán Gasca, and F. Nicolò

1 Technology as Knowledge:
The Case of Modern Engineering Systems

In recent years scholars coming from the fields of history and philosophy of science and technology have devoted much attention to the problem of "technology
as knowledge" and to the emergence of an autonomous *engineering science* in the
Industrial Age[1]. This interest echoes a growing awareness among engineers of the
independence of their conceptual approach with respect to other forms of knowledge, linked to the consolidation of autonomous academic engineering research in
the 20th century. A careful examination of the nature of technological knowledge
appears particularly valuable in view of the pervasive presence of technology in
contemporary life and culture, not only as a result of its impressive achievements,
but through the less obvious influence of its concepts and viewpoints as well.

The activity of engineers and technicians has been traditionally based on the
practical ability to cope with specific situations and to attain the corresponding
specific goal by means of the design and realization of an artifact or structure,
on the basis of past experience handed down by tradition and applied by means
of trial-and-error and rule-of-thumb procedures. But the existence of a theoretical background and of principles underpinning this activity can be traced back
to classical antiquity. The role of geometry and mechanics in surveying or in the
design of machines and structures was explicitly pointed out by Renaissance
artist-engineers. The so-called mixed mathematics formed the core of what was
later to be considered the *sciences de l'ingénieur*, trigonometry and geometry,
applied mechanics, and hydrodynamics. The French scholars and engineers who
taught at the École Polytechnique sanctioned the principle that scientific subjects,
mathematics, as well as experimental natural science, represented a suitable basis
for the education of the modern engineer and for the development of a new theoretical, systematic approach to technical problems, as defined by the word "technology". The merging of the technical engineering tradition with the new scientific
knowledge led to the development of mechanical engineering and to the birth of
electrical and chemical engineering: this process was wrapped up by a "rhetoric of
technology as applied science", which was instrumental in the engineering disciplines obtaining an academic "scientific" status[2].

The growth of industrial engineering and the development of the large modern
technological systems has opened up an entirely new phase in this evolution. The
analysis of industrial production systems and the deployment of the large scale systems typical of post-World War II technology have led engineers to take into con-

sideration not only machines and physical equipment and tools but also abstract objects such as "information", "decisions", "quality". The emphasis on process technology, that is, the planning and control of physical processes as well as information flows, has grown with the evolution of mass production methods, automation, control, communications, and information technologies, and has become a characteristic of modern engineering. The core of such technological developments is theoretical, mathematical based research following new approaches such as those typical of operations research and systems engineering.

2 Scientific Planning and Control: From Manufacturing to Organizational Systems

The Industrial Age has entailed the development of engineering systems, which have replaced single devices and machines as the main concern of the engineering sciences. The archetype of this kind of systems is the factory as a production system: the contribution of several kinds of technology, the interaction between man and machine, the role of a great variety of economic and logistic internal and external conditions are all aspects that render the intuitive operative knowledge typical of the traditional workshop inadequate. In his work *On the economy of machinery and manufactures* (1832) Charles Babbage made a pioneering contribution to the study of these new conditions of manufacturing, considering the elements of productions systems, the classification of different kinds of work, resources allocation and time management. At the beginning of the 20th century the development of mass production was accompanied by a spate of investigations on the scientific planning and control of production, with fundamental contributions by authors such as Frederick W. Taylor and Henri Fayol, and many others by authors working in France, Germany, the USA and the USSR.

The importance given to the idea of building up a scientific, rational approach to management stemmed from the prevailing attitude of confidence in science, the "scientism" which was widespread among the learned society around the turn of the century. The model of science as a body of knowledge and as a method of investigation worked in several ways. An important role was played by the idea of testing and experiment, and the emphasis on systematic analysis and measuring regarding operations and individual jobs in the factory. Also scientific psychology, and specially the behavioral approach, was applied to the study and control of human factors insofar as they influenced productivity. And the idea of introducing mathematical methods into this context began to make headway with the introduction of mathematical programming and the techniques of statistical quality control in manufacturing.

Early studies shared a common interest in understanding the entire production process, rather than its isolated parts, as a way to improve productivity. Thus, specific interdisciplinary methods were searched for in order to support the management of processes, that is, the planning, organizing, decision making procedures, and control of the whole system. The kind of knowledge to be developed was "technological", in a sense that what was looked for was a specific "system technology" ready to be applied in different production contexts. Many other examples of this kind of organizational problem were provided by military operations in World War II. Organizational aspects were an important component in the large defense systems set up in the USA after the war. The development of operations research and mathematical programming provided an entirely new set of mathematical tools for solving decision-making problems in this technological context, which were progressively introduced in industry after the war.

In the second half of the 20th century, the development of automation and the introduction of information technology led to the diffusion of systems for the production of goods and services with a high degree of complexity. The development of mathematical programming and optimization went hand in hand with the development of mathematical control theory and with information science and technology. Management science and industrial engineering (sometimes considered as an equivalent to operations research in the English speaking world) stand at the crossroads of engineering, applied mathematics, and the human and social sciences. Engineering studies concentrate on the example of production systems. But the role of rational decision and the problems of activity analysis of production and allocation in the general context of economics, from a theoretical or applied point of view, began to be recognized in the pioneering contributions of Leonid Kantorovich in the Soviet Union and George Dantzig, Tjalling Koopmans and John von Neumann in the United States, and attracted considerable attention in the second half of the 20th century. Thus, results in the mathematical field of optimization were applied to economics problems as well as to decision problems in management science.

Modern engineering systems have proved to be an interesting source of technological-based concepts, theoretical issues, and metaphors[3]. The development of automation, control, computing, and optimization has stimulated many attempts to transfer ideas from the technological to the biological, social or human context, in which the "system" replaces the classical technical image of the "machine". Norbert Wiener's cybernetic group is one outstanding example of this intellectual trend in Western culture. Herbert A. Simon, Ludwig von Bertalanffy, Walter Buckley are other well-known representatives. In Eastern Europe, the phase of rejection of cybernetics was followed by the blossoming of research in a broadly conceived systems science, regarding the general concepts, theories, methods and techniques associated with analysis, modeling, control and management in various (technological, economic, ecological, social) systems.

This kind of studies followed different patterns in different cultural contexts and was cultivated with varying intensity and many cross-cultural influences in the 20th century. Thus, in recent years, the development of mathematical decision-making, planning and control has led to the consideration of general organizational systems (production systems and other technological systems are only one example); but the idea of building an organization science had been envisioned at the beginning of the century by Bogdanov (Aleksandr A. Malinovskij). In this evolution, an important role was played by the mathematical theories developed for controlling and managing modern technological systems, from the development of radio engineering to spacecraft and missile projects. Mathematical models were a vehicle for conveying technological concepts to several branches of engineering as well as to a broader cultural context.

3 The Interaction Between Mathematical Models and Technological Concepts in 20th Century Technology

Recent studies questioning the view of technology as a mere application of scientific discoveries have concentrated on the comparison between technological knowledge and natural sciences. But what about mathematics, which has been historically considered as essential (at least as a desideratum) for the development of a systematic approach to technical problems? In the first essay of this book, Eberhard Knobloch describes the role of the image and the prestige of mathematics in the passage from ancient to preindustrial technology and machines; and he also explains the limits of its effective role. The importance of this question of image was so great that, as Ivor Grattan-Guinness has underlined, it explains the "Western practice, maintained from the seventeenth until well into the nineteenth century, of calling the professional mathematician a 'geometer', to be distinguished from a mere practising 'mathematician'" (Grattan-Guinness 1994a: vol. 1, 171). Knobloch also gives us an early example (1586) regarding the organizational aspect of the work of an engineer, that is the removal of the Vatican obelisk under the direction of Domenico Fontana.

Several recent historical studies have analyzed the discussions on the role of mathematics in the period of "invention" of the modern engineer and the creation of a new form of "technological rationality" between the 18th and the early 19th centuries; the place of mathematics in the École Polytechnique; and many aspects of the interaction between mathematical physics or geometry and technological knowledge. The relationship of mathematics to engineering knowledge was viewed as a crucial issue at the turn of the 19th century, as is shown in the discussions that took place in that period on the mathematical training of engineers, which involved the relationship between mathematicians and engineers both teaching in technical education institutes. The background to this debate was the professionalization of mathematics – including the creation of an institutional framework and a great development of abstract, pure mathematical research – and the emergence of academic research in the engineering sciences.

A renewal of the classical relationship between technology and mathematics lies behind the definition of the modern concept of engineering system. The figure of Balthazar van der Pol is particularly representative of this, as Giorgio Israel shows in his essay. In particular, mathematical modeling as a distinctive method of 20th century applied mathematics has been tuned up in the technological context, and in the work by van der Pol the idea of mathematical analogy, typical of mathematical modeling, makes one of its early appearances. The technological context here is radio engineering, and the key idea is that of feedback, which will be central to control engineering and thus in the developments considered in the second part of the book.

The other two essays in this first part directly regard the emergence of planning and control mathematical techniques typical of modern engineering systems. The essay by Ana Millán Gasca considers this issue in the context of the history of the applications of mathematics to the non-physical sciences, showing how the early ideas concerning the management of operations and production using quantitative techniques are related to the idea of developing a *mathématique sociale*. In fact, 19th century contributions to mathematical economics by engineers were often developed starting from problems originating in the technological sphere and from the activities and responsibilities of engineers. The reasons for the rejection of the use of mathematics in economics worked also as far as the introduction of mathematics in management and decision-making was concerned: the dominant opinion in the century was that the freedom of human beings could not be reduced to mathematical equations. The essay by Amy Dahan and Dominque Pestre describes the turning point represented in the culture of control by the 1940s and 1950s, thanks to the development of the innovative mathematical techniques of operations research and system analysis and to the extension of their application from war and industrial management to other domains of intervention. This is an important aspect of the interaction of science and technology and of the formation of the scientific-military-industrial complex during World War II and the Cold War, though less well known than the developments in the physical sciences that led to the development of weapons, instruments and devices for defense or civil applications. Both aspects had a strong impact on the material and cultural conditions of life in the second half of the 20th century. In particular, the views of engineers concerning organization and management formed the basis of the idea of a scientific mode of social intervention based on optimization techniques, which were being developed in that period.

The second part of the book focuses on the evolution of several basic concepts at work in the design of 20th engineering systems, considering both control engineering and systems management aspects. In the five essays in this part several historical phases are considered, as well as different technological contexts (military applications, industrial production, decision support). All of them explore the distinctive approach of the "sciences of the artificial" in the age of automatic computerized control systems, with the emphasis on the interaction between technological thinking and mathematical models.

Stuart Bennett presents an analysis of the evolution of control engineering which highlights the development of ideas through different technologies (from the 19th century regulators to servomechanisms, from communications technology to management engineering) and the mutual influences of different mathematical instruments. The idea of feedback is picked up in the essay by Antonio Lepschy and Umberto Viaro as a prominent example of the spread of some key concepts to non-technological contexts; the modern mathematical idea of feedback control in automation systems is also a paradigmatic case of the mathematically formalized 20th century reconsideration of classical issues in the technological thinking. The extension of mathematical control theory to other domains is illustrated also in the essay by Evgenii F. Mishchenko, Alexandr S. Mishchenko and Mikhail I. Zelikin, which reflects also essential characteristics of the Russian approach to control engineering and to the interaction between technological problems and mathematics. The case of environmental sciences is particularly impressive, because it shows the problems of the methodology of mathematical modeling linked to the choice of specific techniques (stochastic, nonlinear, and so on) and to the empirical validation of the models, and the possible consequences of social and economic policies driven by mathematical analysis carried out in scientifically based institutions.

The question of the methodology of model building, analysis and validation is also central in the essay by Andrzej P. Wierzbicki, which describes some important aspects of the evolution from operations research to a general systems science in the second half of the 20th century. He identifies two distinct approaches, hard versus soft, which reflect the differences in systems thinking in the social sciences and in engineering; he also mentions the differences in cultural approach such as those between the two political-cultural areas in the Cold War or between the Western and the Far Eastern approaches (this latter issue is being explored also from the point of view of the various techniques and views on the management of industrial production, and their cross-cultural international transfer, but deserves further research[4]). The chapter picks up the central issue of the mathematical modeling of rationality and decision processes, showing the present trends of research in this field. The essay by Mario Lucertini deals with complexity, a key concept in the management of production systems or large organizations; most fundamental 20th century contributions to management science and operations research can be seen as attempts to cope with process complexity or system complexity in different organizational contexts. Many concrete industrial examples are introduced, but the point of view chosen is that of a general organization science. Complexity is a central issue in contemporary science, which reflects the effort to overcome reductionism. The technological views on the management of complexity are an important contribution to the general scientific and cultural contemporary debate.

This book is one of the results of an interdisciplinary research project on "The evolution of events, concepts, and models in engineering systems" directed by Mario Lucertini with Ana Millán Gasca as scientific manager[5]. The project was carried out at the "Vito Volterra" Centre directed by Luigi Accardi, with the support of the UNESCO Regional Office for Science and Technology in Europe. The series of lectures "Controlling, managing, organizing. Technological knowledge, mathematical modeling and human sciences: development of ideas in the 20th century" organized by the "Vito Volterra" Centre (with the support of ROSTE-UNESCO) and by the Interdepartmental Research Centre on the Methodology of Science ("La Sapienza" University of Rome), directed by Giorgio Israel, at the Italian National Research Council, during years 2001 and 2002, provided a very important context for discussion. We wish to thank Luca Dell'Aglio, Paolo Freguglia, Judy Klein, and Silvano Tagliagambe who lectured in this series together with many of the authors of this book. Finally, we thank Francesca Patriarca for her help in the organization of the project and of the publication of the book.

Mario Lucertini, Ana Millán Gasca, Fernando Nicolò

Notes

1 See the seminal papers by Layton 1971a, 1974, 1988, and the proceedings of the Burndy Library Conference on "The Interaction of Science and Technology in the Industrial Age" published in the journal *Technology and culture*: (Reingold and Molella 1976), specially Mayr 1976. Some interesting examples of recent contributions in this direction, regarding epistemological and/or institutional aspects are: Seely 1993; Vincenti 1990; and Wise 1985.

2 See Layton 1976; Kline 1995. A survey of historical studies on the development of modern *engineering science* is presented in Channell 1989, who considers the following areas: applied mechanics, thermodynamics and heat transfer, and fluid mechanics.

3 See Levin 1999. On the development of the "systems approach" in the USA. since 1945, see Edwards 1996 and Hughes 1998, two contributions that had identified the significance of the ideas and accomplishments in this context, with particular attention to computers and cognitive science (the first) and to the integration of the organizational and automation components (the second). Recent studies on the development of control engineering and automation and on the origins of operations research can be better developed taking into account this cultural context, as in the volume A. Hughes and T. P. Hughes 2000. On the relationship between a "hard" and a "soft" systems approach, see chapter 8. As to the cultural origins of the second one, see the contributions on the origins of systems thinking in Russia and the work of Aleksandr Bodganov in Biggart, Dudley and King 1998.

4 See Zeitlin and Herrigel 2000 and Boyer et al. 1998.

5 The content of the individual articles is the sole responsibility of their authors.

References

Askin, R. G. and Standrige, C. R., 1993, *Modeling and Analysis of Manufacturing Systems*, New York, Wiley.

Barnes, B., 1982, "The science-technology relationship: A model and a query", *Social Studies of Science*, 12: 166–172.

Bennett, S., 1979 and 1993, *A History of Control Engineering, 1800–1930 and 1930–1950*, London, Peter Peregrinus,

Bertalanffy, L. von, 1968, *General System Theory*, New York, George Braziller.

Biggart, J., Dudley, P. and King, F. (eds.), 1998, *Alexander Bogdanov and the Origins of Systems Thinking in Russia*, Aldershot, Ashgate.

Bjørke, Ø. and Franksen, O. I. (eds.), 1981, *Structures and Operations in Engineering and Management Systems*, Trondheim, Tapir

Booth, A. D., 1960, *Progress in Automation*, vol. 1, London, Butterworths Scientific Publications.

Boyer, R. et al. (eds.), 1998, *Between Imitation and Innovation: The Transfer and Hybridization of Productive Models in the International Automobile Industry*, Oxford-New York, Oxford University Press.

Buckley, W., 1967, *Sociology and Modern Systems Theory*, Englewood Cliffs (N.J.), Prentice-Hall.

Channell, D. F., 1982, "The harmony of theory and practice: The engineering science of W. J. M. Rankine", *Technology and Culture*, 23: 39–52.

Channell, D. F., 1989, *The History of Engineering Science: An Annotated Bibliography*, New York: Garland.

Christensen, D. Ch., 1993, *European Historiography of Technology*, Odense, Odense University Press.

Edwards, P. N., 1996, *The Closed World. Computers and the Politics of Discourse in Cold War America*, Cambridge (Mass.)-London, MIT Press.

Ford, L. R. and Fulkerson, D. R., 1962, *Flows in Networks*, Princeton (N.J.), Princeton University Press.

Fox, R. (ed.), 1996, *Technological Change: Methods and Themes in the History of Technology*, Amsterdam, Harwood Academic Publishers.

Gemelli, G. (ed.), 1994, *Big Culture: Intellectual Cooperation in Large-Scale Cultural and Technical Systems*, Bologna, CLUEB.

Gökalp, I., 1992, "On the analysis of large scale systems", *Science, Technology and Human Values*, 17: 57-78.

Herrigel, G. B., 1994, "Industry as a form of order: A comparison of historical development of the machine tool industries in the United States and Germany", in: *Governing Capitalist Economies: Performance and Control of Economic Sectors* (Hollingsworth, J. R., Schmitter, Ph. C., and Streeck, W., eds.), New York-Oxford, Oxford University Press: 97-128.

Homburg, H., 1978, "Anfänge des Taylorsystems in Deutschland vor dem Ersten Weltkrieg", *Geschichte und Gesellschaft*, 4: 170-194.

Hounshell, D. A., 1984, *From the American System to Mass Production, 1800-1932: The Development of Manufacturing Technology in the United States*, Baltimore, Johns Hopkins University Press.

Hounshell, D., 1997, "The cold war at Rand, and the generation of knowledge", *Historical Studies on the Physical and Biological Sciences*, 27: 237-267.

Hughes, T. P., 1976, "The science-technology interaction: The case of high-voltage power transmission systems", *Technology and Culture*, 17: 646-672.

Hughes, T. P., 1998, *Rescuing Prometheus*, New York, Pantheon Books.

Hughes, A. and Hughes, Th. P. (eds.), 2000, *Systems, Experts, and Computers: The Systems Approach in Management and Engineering, World War I and after*, Cambridge (Mass.), MIT Press.

Johnson, R. A., Newell, W. T. and Vergin, R. C., 1972, *Operations Management: A Systems Concept*, Boston (Mass.), Houghton Mifflin.

Klein, J., 2001, "Post-war economics ad shotgun weddings in control engineering", Mary Baldwin College, preprint.

Klein, J., forthcoming, "Economics for a client: The Case of statistical quality control and sequential analysis", *History of Political Economy*.

Kline, R., 1995, "Constructing 'technology' as 'applied science'. Public rhetoric of scientists and engineers in the United States, 1880-1945", *Isis*, 86: 194-221.

Koopmans, Tjalling C. et al (eds), 1951, *Activity Analysis of Production and Allocation. Proceedings of a Conference*, New York-London, Yale University Press.

Kranakis, E., 1989, "Social determinants of engineering practice: A comparative view of France and America in the nineteenth century", *Social Studies of Science*, 19: 5-70.

Kranzberg, M., Elkana, Y. and Tadmor, Z. (eds), 1989, *Innovation at the Crossroads between Science and Technology*, Technion City, Haifa, The S. Neaman Press.

Krohn, W., Layton, E. T. and Weingart, P., 1978, *The Dynamics of Science and Technology.* Dordrecht, D. Reidel.

Laudan, Rachel (ed), 1984, *The Nature of Technological Knowledge: Are Models of Scientific Change Relevant?*, Dordrecht, D. Reidel.

Layton, E. T., 1971a, "Mirror-image twins: The communities of science and technology in nineteenth century America", *Technology and Culture*, 12: 562–580.

Layton, E. T., 1971b, *The Revolt of Engineers: Social Responsability and the American Engineering Profession*, Baltimore, The Johns Hopkins University Press.

Layton, E. T., 1974, "Technology as knowledge", *Technology and Culture*, 15: 31–41.

Layton, E. T., 1976, "American ideologies of science and engineering", *Technology and Culture*, 17: 688–701.

Layton, E. T., 1988, "Science as a form of action: The role of engineering sciences", *Technology and Culture*, 29: 82–97.

Lecuyer, C., 1992, "The making of a science-based technological university: Karl Compton, James Killian, and the reform of MIT, 1930–1957", *Historical Studies in the Physical Sciences* 23: 153–180.

Lenstra, J. K., Rinnooy Kan, A. H. G. and Schrijver, A. (eds.), 1991, *History of Mathematical Programming. A Collection of Personal Reminiscences*, Amsterdam, CWI/North Holland.

Leslie, S., 1993, *The Cold War and American Science: The Military-Industrial-Academic Complex at MIT and Stanford*, New York, Columbia University Press.

Levin, M. R. (ed.), 1999, *Cultures of Vontrol*, Amsterdam, Harwood.

Lucertini, M. and Telmon, D., 1993, *Le tecnologie di gestione. I processi decisionali nelle organizzazioni integrate*, Milano, Franco Angeli.

Lundgreen, P., 1990, "Engineering Education in Europe and the U. S.A., 1750–1930: The rise to dominance of school culture and the engineering professions", *Annals of Science*, 47: 33–75.

Mayr, O., 1970, *The Origins of Feedback Control*, Cambridge (Mass.), MIT Press.

Mayr, O., 1976, "The science-technology relationship as an historiographic problem", *Technology and Culture*, 17: 663–672.

Mayr, O., Post, R. C. (eds.), 1981, *Yankee Enterprise: The Rise of the American System of Manifactures: A Symposium*, Washington, D.C., Smithsonian Institution Press.

Mitcham, C., 1994, *Thinking Through Technology: The Path between Engineering and Philosophy*, Chicago, Chicago University Press.

Noble, D., 1984, *Forces of Production: A Social History of Industrial Automation*, New York, Alfred A. Knof.

Picon, A., 1992, *L'invention de l'ingénieur moderne: L'École des Ponts et Chaussées, 1747–1851*. Paris: Presses de l'École Nationale des Ponts et Chaussées.

Picon, A., 1996, "Towards a history of technological thought", in: Fox 1996: 37–49.

Rapp, F., 1981, *Analytical Philosophy of Technology*, Dordrecht, D. Reidel.

Reingold, N. and Molella, A. (eds.), 1976, "The interaction of science and technology in the Industrial Age", *Technology and Culture*, 17: 621–724.

Sabel, C. F. and Zeitlin, J. (eds.), 1996, *Worlds of Possibilities: Flexibility and Mass Production in Western Industrialization*, New York-Cambridge, Cambridge University Press.

Seely, B., 1993, "Research, engineering and science in American engineering colleges 1900-1960", *Technology and Culture*, 34: 344-386.

Simon, H. A., 1969, *The Sciences of the Artificial*, Cambridge (Mass.), MIT Press, 1996[3].

Vincenti, W., 1990, *What Engineers Know and How they Know it: Analytical Studies from Aeronautical History*, Baltimore, Johns Hopkins University Press.

Warnecke, H.-J. and Steinhilpeer, R. (eds), 1985, *Flexible Manufacturing Systems*, Berlin-New York.

Wise, G., 1985, "Science and technology", in: *Historical Writing on American Science*, (Kohlstedt, S. G. and Rossiter, M. W., eds.), Osiris, 2nd series, vol. I: 229-246.

Wit, D. de, 1994, *The Shaping of Automation: An historical Analysis of the Interaction between Technology and Organization 1950-1985*, Hilversum, Verloren.

Yin, G. and Zhang, Q., 1996, *Recent Advances in Control and Optimization of Manufacturing Systems*, London-New York.

Yurkovich, S. (ed.), 1996, *The Evolving History of Control*, IEEE Control Systems, 16 (3), special issue on the history of control.

Zeitlin, J. and Herrigel, G. (eds.), 2000, *Americanization and its Limits: Reworking US Technology and Management in Postwar Europe and Japan*, Oxford-New York, Oxford University Press.

****

Just as we were working on the manuscript of this book in collaboration with the various authors, Mario Lucertini passed away. As an author of numerous articles on combinatorial optimization and industrial automation, he made an important contribution to the development of operations research and management engineering in Italy. His scientific work was carried out in the framework of a wide range of cultural interests. He deemed it of great importance to study in depth the concepts and methods of the engineer's knowledge, also with the help of the history of science and technology, as well as to disseminate the words and ideas of engineering among the general public. His work at the *Enciclopedia Italiana* and the research project which gave rise to the present book were two of his main activities in this direction. Any further discussions and future research stimulated by this book among scholars of different disciplines and from different countries, like those represented herein, will be a worthy recognition of his intellectual stature.

Ana Millán Gasca and Fernando Nicolò

Part I
Mathematical Methods
and Technological Thought:
Historical Aspects

1 Mathematical Methods in Preindustrial Technology and Machines

EBERHARD KNOBLOCH

The book *De architectura* by the Roman architect and engineer Vitruvius (first century AD) became known in the 12th century; nearly all the works of Archimedes (3rd century BC) were translated by William of Moerbeke in 1269, from Greek into Latin; and the peripatetic *Mechanica* on mechanical problems was translated into Latin in the early 13th century. But these writings had no influence on the practice of medieval craftsmen who did not understand Latin. Moreover, even Vitruvius did not know mathematical statics, only empirical rules. Arches or vaults did not play a real role in his thinking. Utility had traditionally not been a consideration in science, only in the arts and crafts, while technology had got along quite successfully without any assistance from science (Drake 1976). The physics of Aristotle was designed to explain the causes of things, not to be of use to the engineer, the architect or the builder.

Medieval builders undoubtedly used techniques of design and construction, knowledge of which was lost with the phasing out of Gothic building at the end of the Middle Ages (Shelby 1976). Their design technique of deriving the elevation from the ground-plan was a characteristic use of what might be called the "constructive geometry" of the medieval mason. It was almost entirely non-mathematical. It permitted a great many variations to be played on a basic set of geometrical forms.

The architecture of the Middle Ages was founded on empirical knowledge which was transmitted within the building lodges (*Bauhütten* in Germany). The apprentice had to learn from his master the step-by-step procedures for every design and construction problem. There were no technical secrets or "secrets of the craft". The medieval Gothic cathedrals were built by means of empirical rules, rules of thumb; there were no engineering calculations based on mathematically formulated laws of nature, nor did medieval architects formulate any new fundamental law.

The Middle Ages produced no literature describing the construction of the foundations underlying its cathedrals. It is improbable that the tradition of the craft included written instructions for this kind of work, with a few exceptions (Prager 1968). While arches were pointed, piers and pillars were more graceful and slender than in the Romanesque or Norman style, the weight of stone vaulted ceilings being supported by exterior flying buttresses as well as by the walls and pillars. The great central octagon of Ely Cathedral in England, for example, is one of the most beautiful and original designs to be found in the whole of Gothic architecture. The machines that were used in building such structures were mainly for lifting or shifting material; in the 14th century gearing was developed to levels of great complexity (White 1962).

1.1 Renaissance Architects-Engineers

Builders of the late Middle Ages had available common devices as simple winches, pulleys, wheels and cog-wheels. Technical refinements in hoists appeared in the machines of Filippo Brunelleschi (1377–1446) for the construction of the cupola of the Florentine cathedral Santa Maria del Fiore (1420–1436). For according to Vitruvius' conception of the profession of an architect he was at the same time architect and engineer. Still in the 16th and even in the 17th centuries authors of books on machines like Salomon de Caus, Giovanni Branca, Georg Böckler explicitly mentioned on the title pages of their works that they represented both professions.

Brunelleschi's machines were recorded in architectural sketchbooks of the late 15th century (Scaglia 1966), among others by Buonaccorso Ghiberti, Giuliano da Sangallo, the Anonimo Ingegnere Senese and Leonardo da Vinci. Brunelleschi taught masons and carpenters to understand and to work from architectural and machine drawings during the period between 1418 and 1447, when he first used them instead of wooden models for his buildings, hoists and cranes (Scaglia 1987). Before the craftsmen could understand these mathematically formulated drawings, he had to give them oral instruction and to train them in these new work methods.

Brunelleschi succeeded in applying three main inventions: the hoisting machine, the system of chains, the vaulting without armature. Prager (1950) reconstructed the machine: A large, central gin was built on the ground of the cathedral's octagon, and a rotary crane on top of the tambour. A so-called "platform for oxen" was erected on the ground. On this platform an ox was hitched to a horizontal pole, pulling a small carriage for the driver, and rotating the upper end of a central, vertical shaft. On this shaft there was an endless worm or "screw". The worm drove a worm-gear consisting of 16 individual tooth segments installed on a horizontal shaft. This horizontal shaft was held in bearings. It rotated a large drum to wind up the hoisting rope. Later on, the 1:16 gear ratio was probably abandoned and a 1:20 ratio adopted, that is four teeth per segment.

The total weight of the cupola, without the lantern, can be roughly estimated as 25000 English tons, and the capacity of the hoist as an average of 15 tons per day. The machine operated during the entire construction period for the cupola without greater difficulty: It was an outstanding success though his gears were still shaped without mathematical analysis and automatic reproduction of gear tooth curvatures and profiles. In Brunelleschi's times it was usual to expect that a fair degree of accuracy for the intermeshing of the teeth is produced by the initial operation of the drive mechanism itself, that is the teeth being forced to operate by an extra driving effort, wear off the excess material, gradually reducing the friction and allowing operation with an efficiency such as 75% or better (Prager 1950: 519).

Brunelleschi was the first artist and engineer in whom we see art and technology combined with a search for a scientific basis. He was the main architect of the early Renaissance, one of the great developers of Gothic building and an impor-

Figure 1: Brunelleschi's hoist (from Prager, F. D., 1950, "Brunelleschi's Inventions and the 'Renewal of Roman Masonry Work' ", *Osiris*, 9:508).

tant early figure in the evolution of modern structural design and analysis (Prager and Scaglia 1970).

Debates about the construction of cathedrals dealt mainly with the reinforcements required by the large vaults, and their aesthetic effects. While Gothic buildings prior to Brunelleschi had risen to great heights, none had ever covered a space of such breadth as did the Florentine cupola. Little was known about the statics of cupolas; larger ones of every type had collapsed. Vaults were one of the most important supporting systems of construction engineering, but the first theoretical investigations into their operation date only from the end of the 17th century. Robert Hooke (1635–1703) had solved the problem in 1675 stating that the shape of a flexible hanging cord will, inverted, give the shape of the ideal arch to carry the vertical loads. But only David Gregory (1659–1708) gave the mathematical solution in 1697 discussing the shape of the catenary (Heyman 1994).

Architects writing after Brunelleschi, like Leon Battista Alberti (1404–1472), François Blondel (1618–1686) and Carlo Fontana (1634–1714), only recorded empirical knowledge; there was no theoretical approach to statics (Straub 1964). Alberti, it is true, developed the first theory of the different kinds of vaults, but only from a formal, geometrical point of view. We come back to this issue.

Brunelleschi's technical work was characterized by the connection with mathematics and mechanics (Klemm 1965b). He probably knew the work of Blasius of Parma and earlier mathematicians and physicists through Giovanni dell'Abbaco and was personally acquainted with humanists of his time like Paolo dal Pozzo Toscanelli (1397–1482), in a similar manner like, later on, Mariano Taccola with Mariano Sozzini, Francesco di Giorgio Martini with Petruccio Ubaldini, Leonardo da Vinci with Luca Pacioli and Giorgio Valla. At the same time these relationships demonstrated the new social position of such engineers. Brunelleschi developed the painter's perspective, thus reproducing reality by means of mathematical instruments.

His solution of the cupola construction for the cathedral was based on theoretical considerations. He used the ellipse in his solution, enabling him to omit the conventional but expensive centring armatures (that is, trusswork of timber), as well as Gothic buttressing. Vaulting without armatures, that is the freehand vaulting method, which may be seen as a development of the Gothic "permanent centrings" method, became his most famous invention. He proposed to construct a large vault as a double shell reinforced by modified Gothic buttresses and hollow ribs, combined with tie-rings, which were modified Roman features. "Invisible" chains were also known in antiquity. He demonstrated that the abolition of the Gothic buttresses did not require the use of ugly chains across the vault. Historical information can be found in the last of the major prize awards given to him, the document of August 17, 1423: "To Filippo Brunelleschi, the designer and director of the main cupola, for several inventions which he has worked out and will work out for the Building Commission, and mainly for the new model recently delivered by him to said Commission, based on the large tie chain for said cupola, and for the forthcoming perfection thereof – 100 Florins" (quoted in Prager 1950: 484). Obviously two basic types of chains were involved, those of sandstone and those of oak. The system of chains for the cupola had emerged as a cardinal question in 1366 and had been a matter of debate since then. Prager (1950: 494) put it in the following way: "The static problem was quite simple, when analyzed in the light of present-day knowledge, but the situation was extremely complex in 1423". He concluded from the historical evidence that the idea was originated by Filippo alone, against the bitter opposition of Giovanni d'Ambrogio and Lorenzo Ghiberti, that the method was new, that it went far beyond the rediscovery and development of Roman chain designs, that it was useful and successful under the particular circumstances of the Florentine Dome construction, but that it was not particularly successful in the subsequent development of the building art. His method fell practically into disuse after the master's death: one of the most famous inventions of the modern age was a unique performance, steadily praised but seldom repeated by subsequent generations.

Alberti praised him in *Della pittura* (1436), saying: "Who is so dull or jealous that he would not admire Filippo the architect, in the face of this gigantic building, rising above the vaults of heaven, wide enough to receive in its shade all the people of Tuscany, built without the aid of any trusswork or mass of timber" (quoted in Prager 1950: 458 f.).

Brunelleschi was forced to look for a completely new solution, because no one then could build a rigid wooden armature as required by the known methods, even if unlimited sums of money were spent on it (as Brunelleschi's biographer, Giorgio Vasari, remarked in 1568). The Florentine vault spanned over 55 metres in a height of 54 metres over the ground.

The work of completing the cathedral during Brunelleschi's tenure was something of a proving-ground for engineering inventions. It is difficult to decide whether his inventions were founded on empirical or theoretical knowledge, or on both. Doubtless he did not dispose of a mathematical theory which provided him with the statically stable, geometrical shape. But a mathematical analysis of the problem and mechanical considerations of which we do not know anything for lack of documents, whose existence cannot be contested though, led him to a solution and made him a pioneer of intellectual sagacity, which was based on mathematics. His passion for experimentation in mechanics through practice was certainly not without foundation in theoretical studies.

He probably promoted the rediscovery of worm-gear drive. The possible source of its development may have been the writings of Archimedes, the whole of which had been translated into Latin in the 13th century (Scaglia 1966: 94). Mechanical developments in the design of clocks may have helped to promote rediscovery of worm-gear in hoists. Such a connection is mentioned by Brunelleschi's biographer Antonio Manetti.

The rotary crane on a mast, or hoist-crane, is his invention of the period 1417–1419. Although Gian Poggio (1380–1459) discovered a manuscript of Vitruvius' *De architectura*, it came too late to influence Brunelleschi's planning of mechanical devices. The worm-gear with screw and slider was conceived in Florence as early as 1436. At all events, Brunelleschi's achievements made the first half of the 15th century one of the great periods in technological experimentation and progress, preparing the ground for the contributions of Leonardo da Vinci (Scaglia 1966). But it is not clear how long Brunelleschi's machines remained in use; building techniques may have taken one step back as a result of the Renaissance of Vitruvian mechanics. Much more was said and written about the relatively primitive building methods used by the epigones than about the true, constructive development that led from Brunelleschi, through Galileo's work in the mechanical art of hoisting material to modern methods. Brunelleschi's system of chains is one of the important elements of the arts. It was an outstanding example, in the imitation of which the builders and workers of modern times have developed a better understanding of the strength of material, thus laying one of the foundations of engineering as a science.

The first printed work on construction engineering was Alberti's book *De re aedificatoria* (1485), composed between 1443 and 1452. Written in Latin, it was of little use to the practitional, but would have held some appeal for the humanistically minded patron. It continued and even tried to surpass Vitruvius, whose book on architecture appeared in print for the first time two years after Alberti's work. Alberti discussed new subjects like a new theory of cupola construction and rules for vaults. The numerical proportions he gave for stone bridges were of corse not based on statics: they were derived from mathematically formulated empirical rules. The shape, not the static, were dominated by geometry. The building was an organism that consists of lines and material like a body.

He developed a general esthetics of buildings which was based on the three notions of number, relation, and arrangement, that is, on mathematical presuppositions. Music was the model. Hence mean proportionals, arithmetical, geometrical, and harmonical means played an essential role. Beauty was defined as a kind of correspondence and harmony of the parts so that the whole was built according to a certain number, a certain relation and arrangement.

Yet, Alberti himself was at the same time a scientist and a practitioner. Besides Nicholas Cusanus and Roberto Valturio, he was one of the three key figures in the humanist synthesis of praxis and the constructive arts (Long 1997: 21–24). His admiration for Brunelleschi was based not so much on his design of proportionate buildings as on the engineering triumph represented by the dome of the Florentine cathedral, as we saw. For him, the whole of architecture was based on mechanics. Though he separated architecture from craft practice by means of mathematics and design, he also grounded it in engineering for the improvement of civic life. Knowledge and practical life, including practical mathematics, were closely interrelated. In his Italian text *Ludi matematici* (ca. 1450) he dealt with problems of practical geometry, drawing on ancient, medieval and contemporary writings as well as on his own experience. About 1447 he tried to raise a sunken Roman galley from the ground of Lake Nemi. Only parts, however, were brought to light.

At around the same time, Antonio Francesco Averlino (ca. 1400–ca. 1462) wrote the first Italian vernacular treatise on architecture, which was never printed during his life time. The only sources he cited were Vitruvius and Alberti. He was trained as a goldsmith in Florence, probably in the Ghiberti workshop. From the 16th century did Vitruvius' book enjoy a great vogue. Daniel Barbaro (1513–1570), for example, published his commentary on Vitruvius in 1556 (it was reprinted in 1567 and 1584). He combined humanistic studies with practical technology. He published the first Italian really useful translation of Vitruvius' book. Like the other commentators of the 16th century, he emphasized the importance of mathematics for technology and art, writing also about perspective.

Of special interest was a water-screw or water-conveyor, described by him as a movable screw turning in a fixed casing. When Giuseppe Ceredi (ca. 1520–1570) published his *Tre discorsi sopra il modo d'alzar acque da' luoghi bassi* in 1567, on the method of lifting water from low places, he found that no general rules on the optimum construction of water-screws were available. The device was by no means new – its design and construction were discussed by Vitruvius – but he soon realized that the formula given by Vitruvius (length to central shaft diameter, 16:1) was both pointless and impracticable. Ceredi improved on this ratio, and applied for a patent on the Archimedian water-screw, which was not known in Western Europe in the Middle Ages; the patent was granted in 1566. He may have been the first to advocate in print the building of models to different scales and testing them as simulations as a means of optimizing practical efficiency. The history of technology abounds in examples of this practice, most notably with Leonardo da Vinci (Drake 1976).

1.2 Steps Towards Scientific Technology and Technical Mechanics

New elements of the technological development are already to be found in Konrad Kyeser's (1366–after 1405) *Bellifortis* (1405). He provided the first secure evidence of the crank, the most important single mechanical device next to the wheel: it is the chief means of transforming continuous rotary motion into reciprocating motion, and vice versa. The great influence of the work is proved by the considerable number of copies elaborated in the 15th and 16th centuries. It describes late medieval military techniques thus reproducing the ancient, especially Roman projectile-throwing engines like catapults and balistas, but also new weapons like the trebuchet and cannons. The medieval catapult was usually fitted with an arm that had a hollow or cup at its upper end in which was placed the stone it projected. Still Leonardo da Vinci designed a siege balista in the form of an immense stonebow. In other words, the old, traditional weapons were not replaced once and for all by the new fire weapons or the trebuchet.

The trebuchet was already described by Villard de Honnecourt (13th century). It had always a sling in which to place its missile and which doubled the power of the engine. The weight of a projectile (horses, men, stones, bombs) cast by a trebuchet was governed by the weight of its counterpoise. Provided the engine was of sufficient strength and could be manipulated, there was nearly no limit to its power (Payne-Gallwey 1907). The successful attack or defence of a fortified town often depended on which of the armies engaged had the more powerful balistas, catapults or trebuchets. Hence the Renaissance engineers emphasized their ability to construct such powerful engines whenever they looked for an employment by a mighty patron. They promised to construct new, useful and ingenious machines (Popplow 1998). They were aware of a quantitative relationship between the size of the engines and their efficiency without being able to describe this relationship in precise, mathematical manner, that is, to establish mathematical ballistics. The Venetian physicist Giovanni Fontana (1395–1455) described a compound crank in his notebook entitled *Bellicorum instrumentorum liber* in about 1420. He designed rope drive mechanisms and communicating pipes for pressure lines.

The essential step in exploring the kinetic possibility of crank and connecting-rod was taken in Italy (White 1962). The earliest evidence of such a combined device is due to Mariano Daniello di Jacopo, called Taccola (1381–1453/1458), who produced the series of four books *De ingeniis* between 1427 and the end of his life, and the work *De rebus militaribus* in 1449. He put forward many general rules for military engineering and ideas for the construction of devices in order to defend or to attack castles, villages, for the construction of fortifications or defences of ships or boats, of fire arms, of cranes, lifting gears or hoists, or mills and of water supply. Like Kyeser he described old and modern devices, battering-rams, movable towers, catapults, trebuchets. The lifting gears comprehended a large variety of devices like movable elevators for cannons, stationary levers, wooden supports, crane lorries, seesaw cranes and swing cranes, building cranes, cable winches. Containers sometimes functioned as counterbalance though normally the container was replaced by a second object which was going down when the first object was going up.

Taccola used this double-acting principle in the case of cranes, mills, wells emphasizing the importance of ingenuity: an ingenious invention is more efficient than the force of oxen, as he put it. Yet, such statements remained qualitative without being quantified. But his drawings were copied again and again and were at least partly known to his successors Francesco di Giorgio Martini and Leonardo da Vinci. Regarding mills, Taccola drew a very early example of gears consisting of several levels. The axles of the wheels were parallel. Hence the whole operating system was under the grinding system: The animal mill was operated by two horses with a transmission so that the great horizontal wheel (the star wheel) turned two small wheels which were therefore turned more quickly. There were two grinding stones which probably occurred only in the late Middle Ages. Like in the case of lifting gears, he designed many different contrivances to supply water: bucket elevators, wells, different types of pumps (bellow pumps, press pumps without suction, piston pumps, membrane pumps), pipes, mains, siphons. The system of four press pumps, for example, was operated by a crank shaft. The water turned a water wheel, its axis had a cog-wheel turning a snail gear which moved a millstone.

Taccola was proud of his surname "Archimedes of Siena" which was given to him thanks to his ability as an engineer, not as a mathematician. He was unaware of any of the more refined devices developed in antiquity. The Archimedian spiral, for example, was misconceived by him, though Vitruvius had described it correctly. Indeed, the amount of technical knowledge accumulated and distributed in books and available in Taccola's time was small. He did not mention Euclid, Archimedes or Vitruvius, especially the famous machines of Vitruvius' tenth book; nor did he apparently know the peripatetic *Mechanica* of (Pseudo-)Aristotle. But he did contribute to the gradual re-emergence of rational teaching and accumulation of knowledge.

The 15th century saw the elaboration of theories of crank, connecting-rod and governor. The technicians strove to achieve continuous rotary motion, and showed enthusiasm for the flywheel-crank combination. But progress in the development of machinery came about as a result of studies in scientific technology, such as the mathematical and experimental analysis of gear-tooth profiles, lubricated bearings, and load-bearing posts or beams, as exemplified in particular by Leonardo (see below). This new direction is characterized by the greater attention paid to mathematical considerations. Brunelleschi, Giovanni Fontana and Alberti occupied themselves with science, mathematics and physics. Buonaccorso Ghiberti (1451–1516) and his contemporary Bartolomeo Neroni (d. 1571) studied problems of pure geometry, in the case of Ghiberti for example, before discussing traditional military engineering. Ghiberti wrote his *Zibaldone* (Collection) between 1472 and 1483. It contains a collection of drawings of mechanical apparatus which are copied in a number of anonymous engineering manuals.

Francesco di Giorgio Martini (1439–1501) was well acquainted with Taccola's drawings. His *Trattati di architettura, ingegneria e arte militare* are preserved in several original manuscripts (1465/75, 1489/92). An incomplete manuscript once belonged to Leonardo da Vinci: Leonardo added marginal notes and sketches. Parts of Francesco's treatise dealt with mechanical engineering and technology. The machines are the truly new element among his drawings. He studied them systematically because he wanted to rehearse all possible solutions and to combine all known mechanisms. Among the drawings many mechanical devices were generally attributed to later epochs. There are no less than 50 different types of flour and roller mills, including horizontal wind mills, antedating Jacques Besson and Fausto Veranzio by a hundred years, scar mills, pile-drivers, weight transporting machines as well as all kinds of winches and cranes, roller-bearings, anti-friction devices, mechanical cars like those described by Albrecht Dürer (1471–1528), a great number of pumps and water-lifting devices and an interesting water or mud-lifting machine that can be characterized as the prototype of the centrifugal pump. He described original war-machines, offensive and defensive, including the hydraulic recoil system for guns (Prager 1963). He tried to find the correct relation between the length, diameter and thickness of a gun-barrel, and between powder charge and the weight of a cannonball. He recognized that a water-pipe which is narrowed at its end produces a stronger jet of water. He strove for a mathematical foundation of technology. However, he gave no general law: his simple rules always concerned particular examples.

Later authors like Besson, Veranzio, Agostino Ramelli, Vittorio Zonca, and Jacopo de Strada found ample inspiration in the writings of Francesco. His reflections about the fate of invention prove that many of the machines and mechanisms described must have been devised by Francesco: "In building it is necessary to move great weights, for without machines the force of man is of small avail; likewise there is need to draw water in large quantity and convey it long distances. No less useful and necessary will be the construction of mills, and in some places where little water is available ingenuity must be used to help; in other places where there is no water at all, mills must be constructed that work with the wind or by some other means. [...] I decided not to disclose the nature of some of my machines and instruments. [...] For once an invention is made known not much of a secret is left" (Francesco 1967: vol. II, 550). His drawings are of such technical excellence as only to be surpassed by Leonardo himself. What is more, the technological tradition that started with Francesco found its way, and quite early, to the Far East. His second treatise contains a lengthy tribute to Federico II da Montefeltre. There he praises arithmetic and geometry. Drawing is, he says, necessary to every human work, including invention, the explanation of concepts, and operations such as the military arts. It is related to geometry, arithmetic, perspective all of which are necessary for making things with right reason (Long 1997: 36).

1.3 The Achievements of Leonardo da Vinci

Leonardo da Vinci (1452-1519) was personally acquainted with Francesco di Giorgio, whom he met in Milan in 1491, and knew Taccola's drawings at least to a certain extent. Beyond that Leonardo's library included a manuscript by Francesco, Leonardo himself transcribed and paraphrased large portions of Francesco's text on forts and bombards and copied their illustrations into Madrid Codex II (the two famous Madrid Codices I and II have been rediscovered in 1966).

Leonardo was an engineer in the sense that the expression had in his day: an inventor and builder of *ingegni* (complex machines or simple mechanical devices) of every sort and for every type of operation (Galluzzi 1987: 41). Such an engineer typified Renaissance court life. From Taccola to Leonardo, the tasks and functions of the court engineer were essentially the same. They were all men of humble origin who had learned their profession in the Renaissance workshop. Leonardo's papers constitute the most extensive, detailed documentation of the technology of the Renaissance. From P. Galluzzi's painstaking and thorough inquiries into Leonardo's life and work we know that Leonardo went through a crucial intellectual development. His profound interest in theory and his use of practical mechanics to study theoretical issues was by far not always the same in his life.

Brunelleschi's influence dominated Leonardo's technical apprenticeship. Leonardo made big efforts to familiarize himself with his works and chose him as his model. From 1489 onward, he looked for mathematical instruction. Trying to establish general mechanical principles and to test them in concrete applications, he became aware of the fact that geometry was the basic unifying tool for these studies. In 1497 he met Luca Pacioli (1445-1517), who gave him lessons on Euclid's *Elements*. This was a remarkable turning point in Leonardo's activities as an engineer. He outlined the contents of ambitious theoretical works and explicitly aimed at basing practical applications on general principles. Yet, his practical applications never came to depend completely on his theoretical knowledge (Galluzzi 1987: 76). He brought together two separate traditions of the mechanical arts – one based on Aristotle's mechanics and medieval statics and kinetics, the second involving the explication of machines and the mechanical arts of the 15th century engineer (Long 1997: 39). The "disciple of experience" was immersed in the study of mathematics, of the mechanics of solids and liquids.

He made more efforts than any of his predecessors to experiment, to carry out calculations and to look for general rules whenever he studied technical devices. He was artist, engineer, scientist and philosopher at the same time. He divined the methodological importance of mathematics for natural sciences and technology (though only the Baroque period created mathematical physics which made modern technology possible). For him, therefore, mathematics was the model of all genuine knowledge (Fleckenstein 1965). His respect for mathematics grew when he applied it to solve mechanical problems: "There is no certainty in science where one of the mathematical sciences cannot be applied", he noted.

Originally, he appreciated the importance of experiment: "Before making this case a general rule, test it by experiments two or three times and see if experiments produce the same effect" (quoted in Maschat 1989: 234). Yet, from his first Milanese period onward, he was inclined to advance his proposals solely on the basis of theoretical previsions saying explicitly that he has not tested them, thus relying on "thought experiments". Hence there are remarks like (Madrid Codex I): "And it seems to me, although I have not yet tested it, that it should be of equal power", "I believe that these will be better. I have not tested it, because I believe so." His faith in the theoretical foundations from which the foreseen effects derived made experimental verification useless. Some especially interesting inventions and studies of this first Milanese period should be mentioned: the invention of the automatic spinning machine or more generally speaking his lasting interest in automation, his studies of mechanical clocks, of flight and flying machines. He came to the conviction that a flying animal like a bird is an instrument which operates according to mathematical laws, "which instrument it is in the power of man to imitate".

He assigned numbers to things measurable: they were added, subtracted and multiplied to tally with observed results. And indeed, wherever he went during his wandering period (1499–1519), he observed and took measurements. The more he was convinced that all knowledge presupposes the mastery of geometry, the more he tended to studies of a theoretical nature. His applied studies from this time were increasingly based upon the natural laws which he formulated step by step. He recognized the universal validity of the law of action and reaction, known today as Newton's third law. He also gained an amazing knowledge of the relation between the volume, temperature and pressure of vapours and gases.

His attempt to derive practical applications from theoretical premises is especially evident in the treatise he was planning to write on water. The sections of hydraulic engineering should be derived from chapters dealing with theoretical questions (Codex Hammer). In Madrid Codex I he designed mills with a better output and calculated the flow of water from the mouths of canals in relation to the height of the opening. Towards the end of his life he seemed to have seen his "vocation" no longer as that of an engineer, but rather as that of a scientist. According to Benvenuto Cellini (1500-1571), the French king François I called the aged Leonardo "a very great philosopher": "King François, who was extremely taken with his great virtues, took so much pleasure in hearing him speak, that he was separated from him only for a few days out of the year [...] I do not wish to neglect to respect the words I heard the king say of him, which he said to me [...] that he believed that there had never been another man born in the world who knew as much as Leonardo, not so much about sculpture, painting and architecture, as that he was a very great philosopher" (quoted in Galluzzi 1987: 91).

It is true that there might have been colleagues of him who were more capable than he was as military architects, as builders of mechanical devices and machines, as hydraulic engineers. But obviously none of them felt so much obliged to look for a more solid foundation for their activity which remained mainly on procedures and practices acquired in the exercise of their art.

Some technical examples will illustrate the foregoing characterization. The two Madrid Codices I and II are a pattern-book of mechanical engineering. Madrid I belongs to the period 1492–1497; Madrid II was written between 1491 and 1505. They contain some of Leonardo's most brilliant machine assemblies and engineering designs. Madrid I is devoted entirely to mechanics, there is a clear distinction between a section on theoretical mechanics and one on applied mechanics and specific mechanisms. "Mechanical elements" he often alluded to in the Codex Atlanticus is a title under which he intended to collect his reflections on mechanics. Therein, the role of geometrical analysis was crucial for him. He analyzed the principles and criteria of the functioning of machines. He aimed at a systematic analysis of conditions and constructive details that lead towards the rational assembly of useful machines. Thus he laid the foundation for a completely novel general theory of the construction of machines (Maschat 1989). Scaglia's typological index of Leonardo's mechanisms and machines and machines comprehend more than 200 entries (Scaglia 1987).

The idea of a "dissection" of machine was very clear in Leonardo's mind, so that technological drawings played an essential role. He said: "All such instruments will generally be presented without their armatures or other structures that might hinder the view of those who will study them. These same armatures shall then be described with the aid of lines, after which we shall describe the levers by themselves, then the strength of the supports" (Madrid Codex I, f. 82r; Galluzzi 1987: 98). Thus he subjected mechanical applications to rigorous quantitative analysis and geometrical schematization. He tried to enlarge the boundaries of the science of mechanics to include real machines that operated with physical structures, encountered friction and obstacles in the medium, or made use of materials of limited resistance. It is worth mentioning that his anatomical studies depended on mechanical models, that he often resorted to geometrical schematization of the functions of organs.

He had clear ideas about the virtues and limitations of water power. In the Codex Atlanticus, which contains a lifetime's note, he wrote: "Falling water will raise as much weight as its own, adding the weight of its percussion [...] But you have to deduce from the power of instrument what is lost by friction in the bearings". This remark can be interpreted as the first formulation of the basic definition of potential energy. Advice is offered to the engineer for correcting the theoretical efficiency, by taking into account energy losses caused by friction.

He studied intensively problems related to friction in gearwheel trains and in bearings. He recognized the faults of the devices of his day, proposing solutions that have been adopted only in recent times. This applies to worm-gears, screw-jacks, ball-bearings and disc bearings. He suggested, for example, anti-friction devices based on bodies rolling between the working surfaces of a bearing. He studied the transmission of power and movement in belt drives, chain drives and sprocket-wheels. Experiments with perpetual-motion machines were certainly one of the reasons for the great and even increasing interest in friction and in methods of reducing it. His attitude towards all these attempts was critical. He said (Madrid Codex I, f. 148r): "O speculator about perpetual motion, how many vain chimeras have you created in the like quest? Go and take your place with the seekers after gold".

He was aware of the relation of the speed of work to its magnitude, as can be concluded from his power-analysis diagram. In it, the power derived from a hydroturbine is associated schematically with each multiplication of mechanical advantage as the power is transmitted from a large turbine to a small worm-gear acting on a large gear. The faster-revolving worm created in the slower-revolving gear an increment in power in reverse proportion to its speed: "The movement of the pinion and the surface of the wheel is as much more rapid than that of its axis as the circumference of the pinion is contained in the circumference of the wheel". Many of his drawings show for the first time a reaction turbine based on an inverted Archimedean screw, thus preceding by a hundred years Giovanni Branca, to whom the invention of those driving spirals is generally attributed.

He also analysed the efficiency of a treadmill. Simon Stevin (1548–1620), who was the first scholar after Leonardo to analyse mathematically this efficiency, arrived at the same conclusion. In a similar way he made an intensive study of elevating mechanisms like pulley blicks, toothed wheels, springs, screws and nuts, rope-and-chain gearings and wedges, and deduced lasting results in technical mechanics and mechanical engineering.

1.4 Engineers of the 16th Century

The engineers of the 16th century continued the mathematization of technical problems, thus preparing the way for the scientifically applied technology of the 18th century. In general they were the first to try to develop their work scientifically. While Niccolò Tartaglia (1499/1500–1557), Gerolamo Cardano (1501–1576), Giovanni Battista Benedetti (1530–1590) were interested in the practical aspects of mechanics, a second group consisting of Federico Commandino (1509–1575), Guidobaldo del Monte (1545–1607), Bernardino Baldi (1553–1617) was mainly interested in the works of classical antiquity and in the rigorous application of mathematics to mechanics (Drake and Drabkin 1969: 13).

The self-taught theorist Tartaglia published his "newly discovered invention, most useful for every theoretical mathematician, bombardier, and others, entitled New science" in 1537. His *Nova scientia* was inspired by a practical problem of gunnery (Groetsch 1996), he began to mathematize ballistics saying in the letter of dedication to Francesco Maria della Rovere, the Duke of Urbino: "I knew that a cannon could strike in the same place with two different elevations or aimings, I found the way of bringing this about, a thing not heard of and not thought by any other, ancient or modern". Only a century later, Galileo deduced correctly that the trajectory of a projectile in a non-resistive medium is parabolic. But surprisingly, Tartaglia had already noticed that the inverse problem of determining elevation angles leading to a given submaximal range has precisely two solutions. Cardano was like the other northern Italians interested in machines. Among those described in *De subtilitate* (Nuremberg 1550) are the siphon, furnaces, and the Archimedean screw, that is, devices suggested by Hero of Alexandria (1st century AD).

Guidobaldo, mathematician, physicist and chief of staff of the Toscanian fortifications, revived the idea of a new mechanics and was confident that his mathematical reasoning would be confirmed by experimental test. In the preface to his *Mechanicorum Liber* (Pesaro 1577), he wrote (Drake and Drabkin 1969: 241 f): "The origin of mechanics is, on one side, geometry and, on the other side, physics. [...] Whatever helps manual workers, builders, carriers, farmers, sailors, and many others (in opposition to the laws of nature) – all this is the province of mechanics. [...] Also in the war, in the building of ramparts, in fighting at close quarters, in attacking and defending places, there are almost infinite uses of mechanics". Indeed, the introduction of cannon made it necessary to recast the whole system of fortification. The design of these works required the services of a large number of men possessed of mathematical skill. The pursuit of mathematics would lead to technological, and therefore to military and economic supremacy. At the Venetian Arsenal, all the skilled men had a good grounding in mathematics (Keller 1975: 16).

Hence there was greater interest in mathematics and in its applications in the 16th century than before, so that a new profession appeared: the teacher of mathematics who might also survey lands, design fortifications, invent new mathematical instruments and draw up maps and plans. He taught the mathematical side of architecture, cartography, and navigation. Models were regularly used to demonstrate the virtues of particular machines. Guidobaldo seems to be the first to propose to build a model specifically for experimental purposes. Automata were an example of the mathematician's role in technology. This applies to planetaria resembling Archimedes', to cathedral and municipal clocks as great pieces of machinery, to the mechanical toys of the Cremonese engineer Juanelo Turriano made for Charles X.

Thus, there arose in France a genre of applied mathematical literature, the illustrated book of mechanical inventions: Ambroise Bachot, Jacques Besson, Joseph Boillot, Jean Errard (1554–1610), Agostino Ramelli (ca. 1531–after 1608). Of all these books that of Besson, *Theatrum instrumentorum et machinarum* (Lyon 1569), is historically the most important. On the title page he is called "most ingenious mathematician". Richard Eden said about him (Keller 1972: 19): "We may think that the soul of Archimedes was revived in Besson, that excellent geometer of our time". His book was a collection of functioning instruments, objects of fantasy, scientific illustrative material and gadgets. Many machine books followed, like those of Ramelli, Vittorio Zonca (fl. 1616–1627), Fausto Veranzio (1551–1617), Giovanni Branca (1571–1645), Salomon de Caus (end of 16th c.), Heinrich Zeising (16th/17th c.), Jacopo de Strada (1507–1588). Many of them referred explicitly to Archimedes on the title page, for example de Caus, Strada, Branca.

Yet, one must not overestimate the praise of mathematics which can be found in such works. It might be rather rhetoric than the starting point of scientifically conceived machines. We find such rhetoric in Ramelli's *Le diverse et artificiose machine*, which appeared in Paris in 1588. Ramelli was a famous engineer when

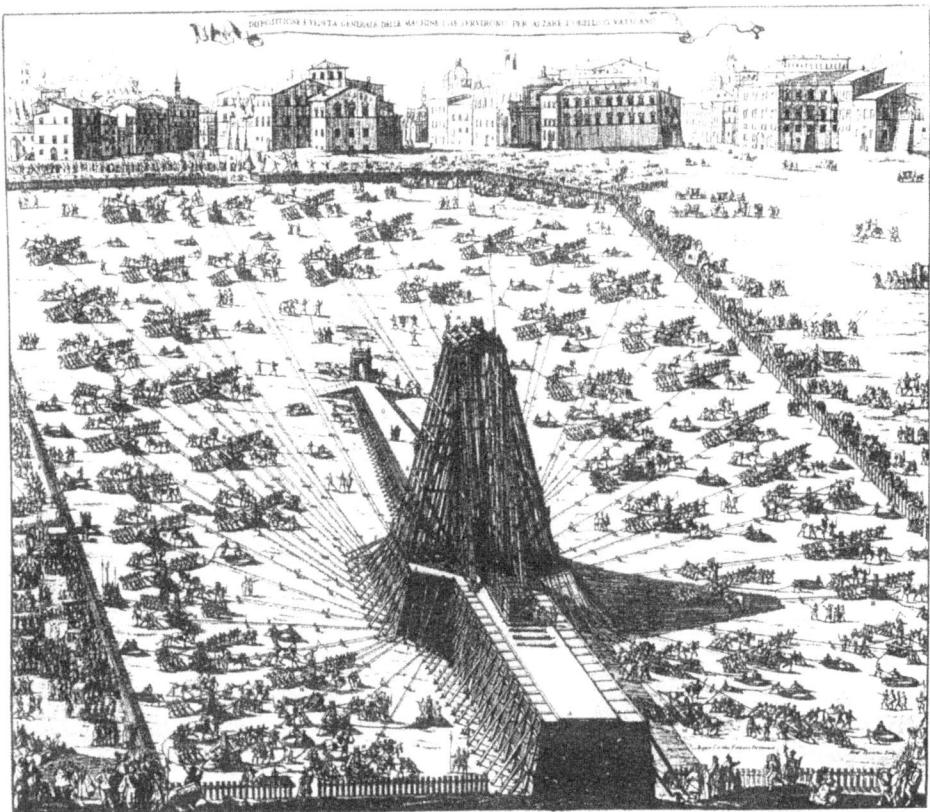

Figure 2: Carlo Fontana's copperplate of Domenico Fontana's hoisting of the Vatican obelisc in Piazza San Pietro, Rome, on September 10, 1586 (from Fontana, D., 1590, *Del modo tenuto nel trasportare l'obelisco Vaticano e delle fabriche fatte da nostro Signore Sisto V*, reprint Düsseldorf, Werner Verlag, 1987, vol. "Übersetzung und Komentare": 73)

he became forty years old. Being young he had participated in the military expeditions of the marquis of Marignano, Gian Giacomo Medici, and had intensively studied mathematics. When he wrote the preface of his work, he could rely on the mathematicians of the Urbino school, that is on "the restorer of mathematics" Federico Commandino and his pupil, Guidobaldo dal Monte (Carugo 1991). When he speaks about the divinity of the mathematical sciences, of the divine Archimedes, when he maintains that the only sure foundation of all liberal and mechanical arts is based on a true comprehension of mathematics, he reminds us, however, of Johannes Regiomontanus' and Luca Pacioli's mathematical divinities. Yet, one must not let oneself be deceived with regard to Ramelli's mechanics considering these enthusiastic remarks. His mechanics did not analyse either the way in which the machines function relying on a mathematical basis. He himself did not realize them either taking into account frictions and appropriate dimensions in the sense of modern engineering. His fantastic catapult and its gigantic dimensions might be a good example in order to illustrate this attitude.

The example allows us to understand what authors like Ramelli meant when they emphasized the importance of mathematics defined as the science of quantities: above all the quantitative aspect and not the functional aspect of the construction, of the machines. Geometry describes the figures and teaches to measure heights, angles, weights etc. Indeed, already Giovanni Fontana, Taccola, and Francesco di Giorgio Martini had designed many instruments and measuring methods (Long 1997: 11; Galluzzi 1995: 122, 130–133). Domenico Fontana (1543–1607) became famous in setting up a Roman obelisk. He was not provided with statical calculations, but calculated the obelisk's weight before estimating how many turns of the mechanism would raise it to the vertical by means of pulley blocks. His success was based on an excellent organization and coordination of the necessary works, operations, activities: the pope had given him unlimited authority.

References

Beck, Th., 1899, *Beiträge zur Geschichte des Maschinenbaues*, Berlin, J. Springer.

Bedini, S. A. and Maddison, F. R., 1966, "Mechanical universe, The Astrarium of Giovanni de' Dondi", *Transactions of the American Philosophical Society Held at Philadelphia for Promoting Useful Knowledge*, New Series, 56 (5): 1–69.

Berninger, E., 1997, "Der Übergang zur Renaissance, Italienische Tradition", in: *Europäische Technik im Mittelalter 800–1400, Tradition und Innovation* (Lindgren, U., ed.), 2nd ed., Berlin, Gebr. Mann Verlag: 551–568.

Besson, J., 1578, *Theatrum instrumentorum et machinarum, Théatre des instruments mathématiques et mechaniques*, Lyon, Bartholomeus Vincentius.

Booker, P. J., 1963, *A History of Engineering Drawing*, London, Chatto & Winches.

Carugo, A., 1991, "Nota sulle fonti letterarie di Agostino Ramelli", in: Ramelli, A., *Le diverse et artificiose machine* (Paris 1588), reprint, Milano, Edizioni Il Polifilo: XXXV–XLI.

Deforge, Y., 1981, *Le graphisme technique, Son histoire et son enseignement*, Seyssel, Champs Vallon.

Drake, S., 1976, "An agricultural economist of the late Renaissance", in: Hall and West 1976: 53–73.

Drake, S., and Drabkin, I. E. (eds.), 1969, *Mechanics in Sixteenth-century Italy, Selection from Tartaglia, Benedetti, Guido Ubaldo, & Galileo*, Madison-London, The University of Wisconsin Press.

Feldhaus, F. M., 1959, *Geschichte des technischen Zeichnens*, 2nd ed. by Edmund Schruff, Wilhelmshaven, F. Kuhlmann.

Fleckenstein, J. O., 1965, "Die Einheit von Technik, Forschung und Philosophie im Wissenschaftsideal des Barock", *Technikgeschichte*, 32:19–30.

Fontana, D., 1590, *Del modo tenuto nel trasportare l'obelisco Vaticano e delle fabriche fatte da nostro Signore Sisto V*, Roma, Domenico Basa (reprint Düsseldorf, Werner Verlag, 1987).

Francesco di Giorgio Martini, 1967, *Trattati di architettura ingegneria e arte militare* (Maltese, C., ed.), Milano, Edizione Il Polifilo.

Galluzzi, P. (ed.), 1987, *Leonardo da Vinci: Engineer and Architect*, Montreal, Montreal Museum of Fine Arts.

Galluzzi, P., 1995, *Les ingénieurs de la Renaissance de Brunelleschi à Léonard de Vinci*, Firenze, Giunti.

Groetsch, Ch. W., 1996, "Tartaglia's inverse problem in a resistive medium", *American Mathematical Monthly*, 103: 546–551.

Hall, B. S. and West, D. C. (eds.), 1976, *On Pre-Modern Technology and Science, A Volume of Studies in Honor of Lynn White, Jr.*, Los Angeles, Undena.

Hermann, A. and Schönbeck, Ch. (eds.), 1991, *Technik und Kultur*, vol. 3, *Technik und Wissenschaft*, Düsseldorf, Verein Deutscher Ingenieure.

Heyman, J., 1994, "The theory of structures", in: *Companion Encyclopedia of the History and Philosophy of the Mathematical Sciences* (Grattan-Guinnes, I., ed.), London-New York, Routledge: vol. II, 1034–1043.

Horologium Amicorum Emmanuel Poulle, 1998, L'Astrarium de Giovanni Dondi, Padoue, Bibliothèque capitulaire, ms D. 39, Paris, École des Chartes.

Keller, A. G., 1972, "Mathematical technologies and the growth of the idea of technical progress in the sixteenth century", in: *Science, Medicine and Society in the Renaissance, Essays to honour Walter Pagel* (Debus, A. G., ed.), vol. 1, New York, Science History Publications: 11–27.

Keller, A. G. G., 1975, "Mathematicians, mechanics and experimental machines in northern Italy in the sixteenth century", in: *The Emergence of Science in Western Europe* (Crosland, M., ed.), London-Basingstoke, Macmillan: 15–34.

Klemm, F., 1954, *Technik, Eine Geschichte ihrer Probleme*, Freiburg-München, K. Alber.

Klemm, F., 1962, "Die sieben mechanischen Künste des Mittelalters", *Die BASF*, 12:46–51.

Klemm, F., 1965a, "Das alte technische Schrifttum als Quelle der Technikgeschichte", *Humanismus und Technik*, 10:27–42.

Klemm, F., 1965b, "Die Rolle der Technik in der italienischen Renaissance", *Technikgeschichte*, 32:221–243.

Klemm, F., 1978, "Physik und Technik in Leonardo da Vincis Madrider Manuskripten", *Technikgeschichte*, 45:4–26.

Knobloch, E., 1994, "Mathematical methods in medieval and Renaissance technology, and machines", in: *Companion Encyclopedia of the History and Philosophy of the Mathematical Sciences* (Grattan-Guinness, I., ed.), London-New York, Routledge: vol. I, 251–258.

Long, P. O., 1997, "Power, patronage, and the authorship of ars. From mechanical know-how to mechanical knowledge in the last scribal age", *Isis*, 88:1–41.

Marx, E., 1926, "Bericht über ein Dokument mittelalterlicher Technik", *Beiträge zur Geschichte der Technik und Industrie*, 16: 317–322.

Maschat, H., 1989, *Leonardo da Vinci und die Technik der Renaissance*, München, Profil.

Mittelstraß, J., 1997, "Leonardo und die Leonardo-Welt. Der universale Mensch als Weltbaumeister", *Freiburger Universitätsblätter*, 138: 51–65.

Naholuha, A., 1960, *Kulturgeschichte des technischen Zeichnens*, Wien, Springer.

Payne-Gallwey, R., 1907, *A summary of the history, construction and effects in warfare of the projectile-throwing engines of the ancients, with a treatise on the structure, power and management of Turkish and other Oriental bows of mediaeval and later times*, London-New York, Longmans, Green and co. (reprint, The Projectile-Throwing Engines of the Ancients, London, Kingprint, 1973).

Popplow, M., 1998, *Neu, nützlich und erfindungsreich. Die Idealisierung von Technik in der frühen Neuzeit*, Münster, Waxmann.

Prager, F. D., 1950, "Brunelleschi's inventions and the 'renewal of roman masonry work' ", *Osiris*, 9: 457-554.

Prager, F. D., 1963, "Francesco di Giorgio Martini's treatise on engineering and its plagiarists", *Technology and Culture*, 4: 287-298.

Prager, F. D., 1968, "A manuscript of Taccola, quoting Brunelleschi, on problems of inventors and builders", *Proceedings of the American Philosophical Society*, 112: 131-149.

Prager, F. D. and Scaglia, G., 1970, *Brunelleschi: Studies of his Technology and Inventions*, Cambridge (Mass.), MIT Press.

Presas i Puig, A., 1998, *Praktische Geometrie und Kosmologie am Beispiel der Architektur*, München, Institut für Geschichte der Naturwissenschaften.

Ramelli, A., 1588, *Le diverse et artificiose machine*, Paris, Casa del'autore (Reprint Milano, Edizioni Il Polifilo, 1991).

Reti, L., 1963, "Francesco di Giorgio Martini's Treatise on Engineering and Its Plagiarists", *Technology and Culture*, 4: 287-298.

Scaglia, G., 1966, "Drawings of machines for architecture from the early quattrocento in Italy", *Journal of the Society of Architectural Historians*, 25: 90-114.

Scaglia, G., 1987, "A typology of Leonardo's mechanisms and machines", in: Galluzzi 1987: 145-161.

Schimank, H., 1961, *Der Ingenieur, Entwicklungsweg eines Berufes bis Ende des 19. Jahrhunderts*, Köln, Bund-Verlag.

Schneider, I., 1970, "Die mathematischen Praktiker im See-, Vermessungs- und Wehrwesen vom 15. bis zum 19. Jahrhundert", *Technikgeschichte*, 37: 210-242.

Schönbeck, Ch., 1991, "Renaissance-Naturwissenschaften und Technik zwischen Tradition und Neubeginn", in: Hermann and Schönbeck 1991: 240-268.

Scriba, C. J. and Maurer, B., 1991, "Technik und Mathematik", in: Hermann and Schönbeck 1991: 31-76.

Sellenriek, J., 1987, *Zirkel und Lineal, Kulturgeschichte des konstruktiven Zeichnens*, München, Callwey.

Shelby, L. R., 1976, 'The "secret" of the medieval masons', in: Hall and West 1976: 201-219.

Simms, D. L., 1995, "Archimedes the engineer", *History of Technology*, 17: 45-111.

Stöcklein, A., 1969, *Leitbilder der Technik, Biblische Tradition und technischer Fortschritt*, München, Moos.

Straub, H., 1964, *A History of Civil Engineering: An Outline from Ancient to Modern Times*, Cambridge (Mass.), MIT Press (German ed. 1949, 1964², Die Geschichte der Bauingenieurkunst, Ein Überblick von der Antike bis in die Neuzeit, Basel, Birkhäuser).

Usher, A. P., 1957, "Machines and mechanisms", in: *A History of Technology* (Singer, Ch., Holmyard, E. J., Rupert Hall, A., and Williams, T. I., eds.), vol. 3, *From the Renaissance to the Industrial Revolution c. 1500- c. 1750*, Oxford, Oxford University Press: 324-346.

White, L. Jr., 1962, *Medieval Technology and Social Change*, Oxford, Clarendon Press.

2 Organization and Mathematics: A Look into the Prehistory of Industrial Engineering

Ana Millán Gasca

One intellectual process that has strongly marked the history of the 20th century is bound up with the specific attention paid to the problems of control, management and organization and the attempt to measure oneself with them on the basis of a rational, explicit representation. This representation had to be suitable for making generalizations, possibly through mathematical modeling, and to replace the traditional approach to such questions based on implicit and intuitive knowledge used to resolve them on a case by case basis. In the 20th century problems arose which acted as an urgent stimulus to these developments: first the diffusion of mass production; then the large-scale logistics problems that arose during World War II, which represented a turning point; the management of international equilibria during the Cold War period and, again during the central decades of the century, the development of the great public and private organizations and the strengthening of the need for a government of society and the economy based on science and the advice of experts; lastly, the current problem of controlling and managing the globalization of communications, production, and commerce.

Management and organization problems became a vast new area of mathematical applications in which modern modeling practice was used; this practice, together with the key concept of system (production system, operations system, social system) made it possible to unify the study of problems as different as those listed above. Besides, since many of these problems emerged in the industrial and technological field and refer essentially to system control, that is, to a central engineering issue, they represented a test bed for concepts and methods typical of technological knowledge and its application to problems that lay outside the scope of traditional machinery and "hardware". During the 20th century the engineering world introduced concepts, terms and metaphors (control, feedback, flexibility, integration, complexity) related to the problem of the management and organization of systems, a problem that thus became a fundamental pathway for accessing technological knowledge in the modern cultural discourse.

2.1 A New Branch of Engineering Science

The Engineer as Manager

The bases of the fundamental organization of modern engineering into the various "engineerings" corresponding to the various action sectors were laid down in the 19th century. This represented an evolution linked to technological development, to the progress of the Industrial Revolution and to the institutionalization of the profession of engineer and of the schools of engineering. In addition to the earlier figure of the military engineer, there now stood the civil engineer, the "ponts et chaussées" engineer, and the mining engineer. Industrial engineering diversified into mechanical and electrical engineering. The 20th century saw the consolidation of new branches such as aeronautical, electronic and telecommunications engineering, corresponding to the development of the various "technologies". These technologies, although involving increasingly complex and interrelated theoretical aspects, nevertheless were related to material elements, to "iron and steel", to objects fundamentally conceived of as "engines", the machines of ancient engineering. Even information technology moved its first steps around a machine, the computer. One innovation brought by the 20th century was instead the birth of a new group of technologies of a different nature compared with the preceding ones, the so-called organizational technologies, that is, a set of tools developed for the purpose of analyzing organizational systems in which men and machines interact closely, and to obtain from them behavior assigned in terms of the optimization of performance, efficiency, costs and so on.

The difference referred to resides in the fact that in this action sector, as in information technology, immaterial objects exert a powerful influence: they consist of flows and processes occurring inside the systems described, flows that may themselves also be related to material objects, such as goods manufactured in a given plant, but also abstract, but nevertheless quantifiable, "objects", such as information. The development of these technologies led to the institutionalization of a new branch of engineering: in the United States this sector is referred to using expressions such as "management engineering" or "industrial engineering". In the 19th century, the term "industrial engineering" was used to distinguish between the action sector of civil engineering, in particular in the public sector, from "new" engineering, which grew up to find solutions to the problems of industrial manufacturing[1]. The now widespread use of the expression in the narrow sense to refer to the organization and management aspects of industrial undertakings may be accounted for by the fact that industrial systems producing manufactured goods provide a main field of application of organization technology. In order to understand the current scope of this area of engineering it may be useful to refer to the definition of "industrial engineering" given in a recent edition (1994–98) of the *Encyclopaedia Britannica*:

> Application of engineering principles and techniques of scientific management to the maintenance of a high level of productivity at optimum cost in

industrial enterprises [...] The industrial, or management, engineer draws upon the fields of systems engineering, management science, operations research, and human-factors engineering. Among his responsibilities are the selection of tools and materials for production that are most efficient and least costly to the company. The industrial engineer may also determine the sequence of production and the design of plant facilities or factories. [...] Many of the decisions of an industrial engineer draw upon management expertise. For example, it is this engineer who conducts time-and-motion studies, who determines wage scales based on an assessment of job skills, and who institutes the quality-control procedures necessary for production of a competitive product.

Therefore many of the decisions made by the industrial engineer depend on his skills in the field of management and organization, and these skills are what characterize the corresponding professional profile. If we examine the historical evolution of the figure of the engineer, it is quite apparent that the organization element has always been an important component of his activity, which distinguishes him from a simple "mechanic", from the technician or the inventor. Historically, the "engineer" was in the first place a military engineer, whose role was related to the attack, the defence and the fortification of defensive positions, that is, to what may be considered one of the first examples of "operations" in the sense that this term originally had in the expression operations or operational research (which emerged, as is widely known, in the 20th century, and precisely in a military context): it consisted of a set of activities put in place to attain a military objective, and which involved a large number of men, vehicles, supplies and so on. And this for example is the first acceptance of the word "engineer" mentioned in the 19th century *Dictionnaire* of Emile Littré, who in this connection mentions the emblematic figure of Sebastien Le Prestre de Vauban (1633–1707)[2]. Among the other responsibilities subsequently also attributed to this figure, namely the construction of public works, roads and bridges, and the exploitation of mines, there was always an important organizational aspect, in addition to the skills relating to the design and use of machines. These management aspects were an important component of the birth of the *modern* engineer (Picon 1992), through which economic matters penetrated his field of action. It could be said to transversely unify the various activities without regard for the specific technical skills required in the various technical fields. It was extremely pronounced in the engineers' corps serving the state, which played an important administrative role[3]. In the private sector, it was increasingly more frequently found in the work of the engineers in 19th century factories characterized by the introduction of machines and the consequent man-machine interaction, as well as by the organization of the productive process in a single place with centralized power and the consequent needs for coordinating the workers' and machines' activities.

How did the 19th century engineer coped with the management and organization aspects of his work? What specialized knowledge did he need in this field? We know that the consolidation of the profession of engineer took place in parallel with the spread of a schooling culture that replaced the old culture of apprenticeship and experience[4]. The engineer's specialist design activity, although retaining many of the aspects related to individual inspiration and experience and being exercised within the bounds of an expertise based on the circulation of information concerning concrete cases, has begun to rely increasingly on a set of scientific principles (originally focused on applied mechanics, but then gradually extended) and mathematical tools (mathematical analysis, descriptive geometry)[5]. Specialization of the engineer's work led to the development of specific course syllabuses in the various sub-sectors: civil (or civilian), mechanical, chemical, electrical engineering and so on. Management and organization continued instead to be considered practical activities based on high level training (an important element for the authoritativeness of state engineers) and on professional experience. The institutionalization of a sector of engineering studies devoted specifically to this type of task occurred in the 20th century, in parallel with the emergence of the modern figure of the manager, as conceived of by the scientific management movement, and thus involving the development of a series of specialist techniques, partly of a quantitative nature, aimed at assisting or replacing subjective, case by case, evaluation[6].

In the following we shall describe several historic transitions leading to the creation of a systematic, theoretical "engineering" approach to the problems of the administration, organization and management of operations and of activities up to the modern formulation in terms of decision making, with a strongly formalized approach. A fundamental role in these developments was played by the debate on the use of mathematical techniques. In a historical perspective, these developments are closely associated with the outcome of the programme of the mathematization of the non-physical sciences dating back to the Enlightenment period, and which had a general bearing on the economic, biological and social sciences[7].

Mathematics, Systems, Organization, and the Human Factor

We have already mentioned one of the fundamental characteristics of industrial and management engineering, which distinguishes it from the "classical" ones, namely the fact that, although still referring to concrete "objects" and having a material reality (such as a factory), it actually refers fundamentally to abstract objects such as decisions, information, flows and processes. A second characteristic, closely linked to the first one, is that it today presents a high level of mathematization and indeed its very emergence as an autonomous sector is marked by a specific profile that the mathematical techniques contribute strongly to defining. The institutionalization of these studies into a 'respectable' and autonomous academic subject, mainly addressing the analysis of organization and management problems encountered in industrial production in the early 1960s in Britain were described in the following terms[8]:

> Production Engineering as university discipline was then new, and as such needed to gain academic respectability, as well as industrial usefulness. Production Engineering as university subject was contemporary with Management Science, also emerging particularly through Operations Research and

other studies, the tendency being to subject the treatment to mathematical rigours. Production Engineering involved the production processes, product engineering, coordination and control through industrial engineering, economics, etc., and these in turn consisted of a collection of mostly numerical techniques aimed at controlling and improving production and its economy.

Indeed the central decades of the 20th century witnessed the development of new mathematical techniques conceived precisely to tackle the problems of programming and organization (even if linked to such classical mathematical sectors as optimization, probability, graph theory and discrete mathematics in general), namely the techniques of operations research. Although in many cases they did not grow as a direct response to industrial problems, these techniques, tools and models lie at the base of many modern technologies applied to the management of industrial production, so much so that the mathematical techniques and corresponding technological and engineering sectors may be identified, as is emphasized for instance in the entry on "Industrial Engineering and Production Management" in the *Encyclopaedia Britannica* (ed. 1994–98):

In the 1970s and 1980s industrial engineering became a more quantitative and computer-based profession, and operations research techniques were adopted as the core of most industrial engineering academic curricula in both the United States and Europe. Since many of the problems of operations research originate in industrial production systems, it is often difficult to determine where the engineering discipline ends and the more basic scientific discipline begins (operations research is a branch of applied mathematics). Indeed, many academic industrial engineering departments now use the term industrial engineering and operations research or the reverse, further clouding the distinction.

A handbook published in the mid 1980 effectively describes the coexistence between the problems and the conventional tools of industrial engineering (IE), that is, those typical of early 20th century scientific management – process planning, facility design, work measurement, job evaluation – and the approaches and techniques of operations research (OR) – linear programming, optimization methods, statistical quality control, inventory control, queuing analysis, networks flow analysis – (Miller and Schmidt 1984: 8):

Because of the contrast in the mathematical orientation between traditional industrial engineering and operations research, it may be useful to think of these two areas as a spectrum or continuum of activities with traditional IE at one end and OR at the other. Traditional IE tends to be more applicable to problems in a manufacturing environment. On the other end of the spectrum, OR has a broader scope, being oriented towards more macro level problems in a wide variety of application areas of which manufacturing is only one. Operations research also places heavier reliance on mathematical concepts, especially mathematical models, than does traditional IE.

[...] However, there is a great deal of overlap [...] An example of this type of activity is facility design. Industrial engineers have traditionally been concerned with facility design issues such as determining the best layout for a plant or the best location for a distribution warehouse. Traditional analytical tools have been used, such as flow diagrams, templates, and guideline checklists. However, more recently OR techniques such as queueing analysis and mathematical programming have been successfully applied in resolving the same facility design problems. In addition, the scope of facility design has been broadened beyond the industrial context to include layout and location problems for post offices, airports, sports organizations, and other services industries. Other activities and problems areas in the middle of this IE/OR spectrum include inventory control, forecasting, production scheduling, and quality control.

The use of the mathematical models of operations research has led off in the direction of greater generality in the development of this engineering sector, the broad scope of which has become management and organization referring to production systems in the general sense (including manufacturing, distribution, and service operations), as well as other organization contexts. From the terminological point of view, the expression "production and operations management" used in the title of many classical textbooks, or else the expression "management engineering", understood as an engineering specialization, corresponds to this development. This kind of general point of view is influenced by the approach typical of operations research, which has become a fundamental industrial management tool, although the applications potential of which extend outside the narrow field of production. This kind of development is also marked by the modeling approach used in the application of mathematics, which, by means of mathematical analogy, allow connections to be established between apparently distant problems.

A third aspect marking this branch of engineering science is the presence of the "human factor": man – in the first manufacturing plants exclusively the worker, man, woman or child, and later a wide variety of persons performing different roles, and with different responsibilities and technical skills, who interact among themselves and with the machines – has been one of the fundamental factors ever since the first analyses of industrial production were made; indeed it is precisely the human factor that has stimulated many of the pioneering studies in industrial management. Furthermore, this aspect brings this sector of the engineer's knowledge closer to the social sciences; the programme of mathematization has thus had to cope in this environment with the same epistemological obstacles that were encountered in the field of economics and the social sciences relating to the possibility of quantifying events involving the human being and his freedom of choice. On the other hand, the presence of the human factor has not ceased to increase and is not limited to the problems treated by ergonomics and human factors engineering; indeed, the formulation of management and organization problems in terms of decision making has brought the basic issues regarding this field back to the behavioral and cognitive sciences.

In this sense the research of Herbert Simon (1916–2001), who played a leading role (starting from the early investigations in the field of municipal administration, dating back to 1936), was emblematic in the development of operations research related to problems of industrial organization and management. In the early 1950s he worked at the Graduate School of Industrial Administration set up at the Carnegie Institute of Technology at Pittsburg, a newly founded school dedicated to this kind of innovative research. His contributions of this period include studies on inventory control carried out in collaboration with Charles Holt (Klein 1999). However he formulated this research in terms of "organizational decision-making" (starting with his *Administrative behaviour: A study of decision-making processes in administrative organization*, published in 1947), and of "rational behavior". The development of these ideas led him, in a later stage of his work, to his brilliant contributions to the artificial intelligence programme. One second line of approach is represented by the game theory of John von Neumann (1903–1957): developed as a programme of renewal of theoretical economics, it represented an important theoretical point of reference for the development of operations research. However, in von Neumann's thinking game theory was also related to the study of rational decisions, and this kind of problem led him, in the final stages of his activity, to develop the theory of automata, a seminal work in artificial intelligence and in automation.

Lastly, a fourth aspect distinguishing industrial and management engineering is that its development was linked to that of systems engineering, which was developed in the second half of the 20th century to cope with the large technological projects. There were multiple interactions, also through the mediation of mathematical models of systems and control theory. However, it would be necessary to make a more general reference to "systems thinking" in the broad sense, as an intellectual trend that underwent an extraordinary development in the 20th century, and that manifests itself in very different times, places and cultural contexts through different schools of thought. Natural systems, artificial systems, human systems and man-machine systems have been conceived of within it, following very different approaches and with more or less appeal to mathematics. And while systems thinking has lent concepts and methods of thinking about reality to control engineering and industrial engineering, it has itself also been strongly influenced by them. In this sense, technology appears here less than ever as knowledge emerging only from the application of scientific theories but rather as a specific form of knowledge that interacts with other knowledge, leading to the development of an organization science that seeks to be applied far beyond the narrow confines of industrial production systems, but is marked by an engineering approach. This is a significant aspect affecting our understanding of the 20th century cultural history[9].

2.2 Quantitative Studies and Labour and Production Organization: the Early Attempts between the 18th and 19th Centuries

Between the late 18th century and the early 19th century, we find one of the first explicit formulations of the "organization problem" regarding labour and production, and several of the pioneering attempts to introduce quantification into this context. To some extent these ideas emerged as a result of the increasing institutional responsibilities of engineers, especially in France, during the period of the Revolution; however, they were dictated also by the emergence of economic and cost control aspects in their work, in a socioeconomic situation in a state of great flux. While it is true that in this period engineers played a fundamental role in the service of the State and in its defence, the advancing Industrial Revolution and the birth of the idea of market were also to play a part in the radical changes that occurred in the role requested of and the demands made on the engineer. One impressive example of the echo of the reflections on economic policy among engineers concerns Gaspard Riche de Prony (1755–1839), one of the leading engineers of his time and director of the École de Ponts et Chaussées from 1798 until his death. In 1792, summoned by the French government to direct the production of a series of mathematical tables to spread the use of the decimal system, he set up an organization based on the writings of Adam Smith (1723–1790) dedicated to the advantages of the division of labour in the first chapter of his *Inquiry into the nature and causes of the wealth of nations* (1776)[10]. Following the example of pin manufacturing used by Smith to illustrate his ideas, he laid down a division of responsibilities and type of operation that would allow maximum accuracy of calculation and minimize the time required and costs. Oddly enough, also de Prony's teacher, Jean Rodolphe Perronet (1708–1794), who directed of the École de Ponts et Chaussées from its foundation (1747) to his death, and was a leading engineer during the closing years of the *ancien régime*, became interested in pin manufacture[11].

The early ideas of the French engineers, although still highly sporadic, entailed attempts at quantifying technical reality that were typical of engineers working during the Enlightenment period. They shared the confidence in mathematics, considered a rigorous foundation of knowledge, typical of their era, and turned to it in an attempt to provide a theoretical framework that would allow new tools to be identified that could be used to solve the technical problems and thus overcome the limitations of the tools available at the time, which were based essentially on experience rather than on systematic 'rational' analysis. The assumption, which led also to the creation of a centre of higher technical training such as the École Centrale des Travaux Publics, later École Polytechnique (in 1794), was that mathematical and scientific culture would ultimately lead to a new approach to problems by engineers, namely the transition from engineering as an "art" to actual "engineering science", to use the expression adopted as early as 1729 by Bernard Forest de Bélidor, professor of the school of artillery of La Feré.

The descriptive geometry developed by Gaspard Monge (1746–1818) made available a graphic language and a conceptual tool for the treatment and design of spaces. The tool available in the field of analysis, the calculus of variations, seemed perfectly suitable for translating into mathematical terms the prescriptive problems typical of engineering practice. This branch, which was one of the more characteristic products of 18th-century mathematical research, had been invented by Leonhard Euler, and developed by him with the fundamental contribution of Joseph-Louis Lagrange, who used it in his analytical formulation of mechanics. Charles Augustin Coulomb (1736–1806) tested it in the field of applied mechanics, in his best known work in the field of engineering research, *Essai sur l'application des règles de maximis et minimis à quelques problèmes de statique relatifs à l'architecture* (1773), which represented an attempt to proceed beyond obtaining numerical solutions on a case by case basis. The idea of taking one step beyond quantification to attempt actual mathematization was what enlivened also his pioneering studies on efficiency and performance of work from the productivity point of view. These studies were developed in parallel with his work in the Corps du Génie, as early as the long period he spent in Martinique (1764–1772), during which he was involved in the construction of Fort Bourbon, a project in which a large quantity of human resources, materials and expenditure was deployed. In his work *Résultats de plusieurs expériences destinées à déterminer la quantité d'action que les hommes peuvent fournir par leur travail journalier, suivant les différents manières dont ils employent leurs forces* (1799). Coulomb concerned himself with labor allocation, applying the "théorie de maximis et minimis". Another attempt to solve a problem referring to the construction of fortifications by expressing it in terms of optimisation was made by Monge himself (1746–1818), who was working at the time on the problem of minimizing the cost of transporting masses of soil and other materials as a function of possible itineraries and related conditions; for example, the need to cross a river, both in the case there are one or two existing bridges, and in the case it is necessary to build them. In his characteristic mathematician's style, in his *Mémoire sur la théorie des déblais et des remblais*, he addressed the problem from the point of view of abstract geometry, considering it first on a plane and then in space, using differential techniques[12].

These pioneering studies, carried out for a two-fold purpose, ideal and concrete, actualized the idea of the mathematically based 'rationalization' of the activities that came within the scope of the engineer. They reveal that new form of technical rationality in which reality is considered dynamically, and thus processes, flows and time can be identified (Picon 1987–88, 1989, 1996). Moreover, again in France, although in more direct relation with organization problems in industry, in the early 19th century several substantial contributions were made in the field of accounting, as in the *Essai sur la tenue des Livres d'une manufacture* (1817) by Anselme Payen. The transition from commercial accounting, which has a long history behind it, to accounting problems in manufacturing, leads to new factors being considered – mainly costs – and to somehow or other taking into account the problem of the productive process[13].

Nevertheless, it was an English author, Charles Babbage (1792–1871), Lucas professor of mathematics at Cambridge University, who first clearly stated the need to rationalize productive activities and who explicitly proposed making a quantitative approach to the problem in terms of process and optimization. His ideas are contained in a book devoted to manufacturing production entitled *On the economy of machinery and manufactures* (1832)[14]. In view of Babbage's peculiar cultural and intellectual profile, his ideas proved highly original, as they represented the focus of aspects of the French engineering approach, the wealth of experience accruing from nearly half a century of development in English factories and, lastly, the early ideas on political economy expressed by Smith[15]. This book by Babbage is actually linked – as he states explicitly in the concluding chapter – to his attempts to renew the study of mathematics and scientific practice in his country, and to overcome the dichotomy between the University and the Royal Society on the one hand, and the rich English technological and industrial situation on the other. One year earlier, in 1831, he published a fierce attack on the Royal Society, *Reflections on the Decline of Science in England, and on some of its causes*, and participated in the establishment the British Association for the Advancement of Science, a scientific institution the main purpose of which was precisely to serve as a link between science and economic and social factors. Babbage, who during his first journey to Paris in 1814 had met de Prony, whom he venerated, shared the analytical and scientific approach to the technical reality of French engineering of the Enlightenment; and in this book he cites the contributions made by Coulomb, Perronet and de Prony. It is, as he himself writes, a by-product of his studies on automated calculus[16]:

> The present volume may be considered as one of the consequences that have resulted from the calculating engine, the construction of which I have been so long superintending. Having been induced, during the last ten years, to visit a considerable number of workshops and factories, both in England and on the Continent, for the purpose of endeavouring to make myself acquainted with the various resources of mechanical art, I was insensibly led to apply to them those principles of generalization to which my other pursuits had naturally given rise.

The book consists of two parts. The first develops the entry "On the general principles which regulate the application of machinery to manufactures and the mechanical arts" published in 1829 in the *Encyclopedia Metropolitana*. It describes an attempt at classification of the machines used in the industrial field according to their mode of action, above and beyond their uses in different sectors of manufacturing: the detailed examination of many examples is used to identify general principles governing the use of machines. The chapter that concludes this first part, "On the method of observing manufactories", is an anticipation of Babbage's views on factories, which extends beyond the specific mechanical art and approaches the consideration of production as a system. It is illustrated fully in the second part entitled "On the domestic and political economy of manufactures", in which he explains the existence of a type of problem that lies outside the mechanical aspects of the manufacturing of goods[17]:

The *economical principles* which regulate the application of machinery, and which govern the interior of all our great factories, are quite as essential to the prosperity of a great commercial country, as are those mechanical principles, the operation of which has been illustrated in the preceding section.

This new type of problem, no longer related only to mechanics or to the function of machinery, arises out of the changes to the productive process produced by the Industrial Revolution[18]:

> A considerable difference exists between the terms making and manufacturing. The former refers to the production of a *small*, the latter to that of a *very large number of individuals* [...]
>
> If, therefore, the *maker* of an article wishes to become a *manufacturer*, in the more extended use of the term, he must attend to other principles besides those mechanical ones on which the successful execution of his work depends; and he must carefully arrange the whole system of his factory in such a manner, that the article he sells to the public may be produced at as small a cost as possible.

The creation of the manufacture or factory, which replaces the craftsman's workshop or cottage industry, thus represents a quantitative change: an increase in the physical size of the place of production and an increase in production volume. There is also a new external aspect, namely the market and competition. Both aspects make it necessary to "arrange the whole system" in order to minimize the price of the product: Babbage's approach thus contains the idea of management of the system for a purpose or objective that is quantifiable and thus optimizable (in this case, minimizing the cost). He takes account of numerous aspects, such as price, raw materials, over-manufacturing, enquiries previous to commencing any manufactory, quality control or verification. As far as the idea of organizing and quantifying the productive process in the strict sense is concerned, the two chapters dedicated to the division of labour are worthy of note. Babbage continues on from the ideas of Smith and points out the three causes of the advantage resulting from the division of labour listed by the latter, that is, the greater skill acquired by the specialized worker, the reduction in the time wasted between one activity and another and the invention of new machines or their improvement, which lead to increased productivity. However, one further cause must be added, as already mentioned by Melchiorre Gioia (1767–1829) in his *Nuovo prospetto delle scienze economiche* (1815–1817)[19]:

> Now, although all these are important causes, and each has its influence on the result; yet it appears to me, that any explanation of the cheapness of manufactured articles, as consequent upon the division of labour, would be incomplete if the following principle were omitted to be stated.
>
> That the master manufacturer, by dividing the work to be executed into different processes, each requiring different degrees of skill or of force, can purchase exactly that precise quantity of both which is necessary for each process;

whereas, if the whole work were executed by one workman, that person must possess sufficient skill to perform the most difficult, and sufficient strength to execute the most laborious, of the operations into which the art is divided.

He thus identifies in the division of labour the organizing principle that "creates" the system (the overall performance of which is determined by the interaction among the various elements or functional units). It opens the possibility of quantifying, and thus of providing the manager of the plant with precise indications that enables him not only to increase productivity but also to lower costs (there is thus an implicit idea of decision-making by a "master manufacturer"). To demonstrate this he gives a "numerical example" which again refers to the manufacture of pins, used as an example by Smith, who had observed the process directly in a small workshop employing ten workers. Babbage considers the (manual) production process, divided into 7 operations (from wire-drawing, to punching the paper on which the pins are presented), for which he identifies the time for execution, the worker's daily earnings and the price of the operation), which involves 10 workers possessing different levels of skill. The quantities handled are listed in a table, and he supplies a second table using Perronet's data[20].

In the chapter cited on the methods used to conduct an enquiry in a manufactory, he referred to the qualitative problem of identifying processes and single operations, as well as to the quantitative problem of measuring human work in terms of 'quantity considered a fair day's work", of "number of operations", of rates and speed. In this connection, he cited Coulomb's opinion, pointing out that the French scholar had worked on this kind of observation and measurement[21]. However, rather than actual concrete technical developments, in Babbage's work the problem arises for the first time of organization conceived of as an issue amenable to systematic treatment and to being quantified through the introduction of the principle of the division of labour. He actually considers this principle from a very general point of view, in which goes beyond the specific case of the industrial productive process. This is shown by the fact that, in the chapter "On the division of mental labour", he deals with the organization of labour used by de Prony for his 1792 mathematical tables project[22].

2.3 Rationality and Mathematization

The development of the early studies on political economy between the late 18th and early 19th century was linked to the attempt to understand the market mechanisms (price formation, the idea of utility) from a very broad point of view. The problem at stake was the prosperity and economic progress of a nation or, even more in general, civil coexistence in modern society. But there were also extremely concrete implications, very acutely perceived by Babbage, as far as the operation of the factories and their success against the competition were concerned. The example of pin manufacture symbolically represents this link as it consists of a small-scale but material achievement of the division of labour inside society. Smith's ideas were taken up by Babbage (and by de Prony), although the point of view were changed quite significantly. Indeed, in Smith, the division of labour is not the result of conscious reflection on the merits of this form of organization, but of the gradual progress of history; it is one of the manifestations of the "invisible hand", that is, of the fact that collective prosperity progresses as a result of behaviors dictated by the quest for individual advantage, in this case leading to an increase in productivity. For Babbage and de Prony it is rather a tool in the hands of man that can be used to achieve a goal, namely the economy of time and costs. In the case of the industrial entrepreneur, to minimize the price and, in the case of the organizer working on behalf of the government, to minimize expenses, although in both cases retaining the quality of the end product. This different approach is certainly the result of the different way an engineer behaves, who is always seeking to attain an objective and to exert control (over the trajectory of a projectile, water flow, the working of a machine). Here this attitude of mind was extended to cover a new kind of problem involving organization in the presence of a first degree of "complexity" and thus of a certain degree of uncertainty concerning the behaviour of the *system*. In the new manufacturing plants a whole new series of problems had to be faced which were linked to the economic and social conditions, to prices, raw materials, but also to taxes, to workers' protests and so on, and the engineers and the industrial entrepreneurs were the first to have to cope with these problems.

The introduction among French engineers of proto-economic elements[23] was influenced by the approach to political economy and to the idea of a social science typical of the reforming spirit of the Enlightenment, as developed in the economic thinking of Turgot (1727–1781) and in the *mathématique sociale* of Condorcet (1743–1794): for these authors, it was not just a question of describing social reality, but rather of establishing rational rules to govern it. Condorcet's idea of applying mathematical tools to this context, which was cultivated also by the idéologues (to whom also de Prony was close), was naturally accepted by the engineers, who shared the confidence in these tools typical of the period. However, as shown by B. Ingrao and G. Israel (1990), during the period of the institutionalization of economic studies in France starting in the 1830s, the project of the mathematization of economic problems ran into radical opposition to the benefit of a political economy that was remote from abstract models and more concrete, empirical and historical. This abandonment is linked to that of the idea of an active intervention in the government of the economy and of society, in favor of a descriptive economics and of rigorously laissez-faire policies. As far as the world of engineers is concerned, in the analysis made by A. Picon, the discussions of the commission set up in 1830 to reorganize the Ecole des Ponts et Chaussées are an obvious sign of a change linked also to the equilibrium between art and science (and mathematics) in the engineer's knowledge (Picon 1989: 168, my translation):

> Visible as early as 1825–1830 in the sector of civil engineering and building, the first symptoms of industrialisation thus seem to run counter to the constitution of a body of unitary knowledge.
> Faced with such a situation, the most realistic attitude is to let oneself be guided by the applications, while favouring an approach in terms of produc-

tion process and line, production process for the works and technical-economic lines, the mastery of which is of importance in the first instance to the engineer. To achieve this mastery, the majority of the members of the Commission deem it of interest to set up courses on administrative law, accounting and political economy. [...] In this context, the status of mathematicians can only become increasingly instrumental. At the same time as their use in the building arts becomes more general, they lose part of their authority. [...] What it is attempted to establish is a global knowledge of technical processes that will allow the ancient conflict between science and the engineer's art to be resolved, a knowledge that one is tempted to define as 'technological', to use a term employed by the 1930 Commission in its conclusions.

Among the engineers, under the pressure of practical problems and of demands for action in their professional activity, both in the public and the private sector, there nevertheless continued to be a traditional greater receptiveness to mathematical studies regarding the management aspects and economic factors related to their activity. This interest may be accounted for by bearing in mind their strong mathematical background and the increasing role played by mathematical tools and language in engineering science. The historiography of economic thinking has long since identified a French tradition in economic studies from engineers. One of the earliest representatives is Achylle-Nicholas Isnard (1749–1803), a member of the Ponts et Chaussées corps and author of *Traité des richesses* (1781), as well as a forerunner of Léon Walras (1834–1910), the founder of general economic equilibrium theory. The influence of 19th century French engineering and of the Paris school of engineers accounts for the widespread nature of this tradition of economic studies developed by engineers in many countries, which were often related to management problems (Teocharis 1994).

– Indeed, as P. Lundgreen has emphasized, the 19th century consolidation of the French system for training engineers in the École Polytechnique and the écoles d'application, produced an engineer who was tilted much more towards management and organization activities than towards specialized technical design. In this context, mathematics played the same role that in other elite-training systems as that played by classical studies and training in law; and this profile, typical of a state engineer's training, was extended in the latter half of the century to the industrial engineers of the École Centrale des Arts et Manufactures (Lundgreen 1990, Shinn 1980, Weiss 1982). Mathematical training represents one of the characteristic features of French engineering as a profession and had thus an important role in the establishment of one of the most influential technocratic models. It provided a guarantee of intellectual rigour, which allowed engineers to engage in all the quantitative aspects of their function, not only as a technical activity, but also in its accounting, economic and organizational aspects. However, in addition to numbers, great importance was attributed to the discretionary powers of the engineer, who was capable of making pondered decisions that extended beyond the merely quantifiable aspects (Porter 1995). The attempts to apply mathematics to the decision-making and management aspects of this professional activity thus clashed with a concept of rationality that refused to accept to be reduced to mathematical rules.

Babbage's project, which referred to the management of manufacturing production although its general approach allowed it to be extended to other aspects of organization, had only a small following in the 19th century. His prestige no doubt contributed to the dissemination of also his *On the economy of machines and manufactures*, of which four editions were published in 1832–1835 as well as two reprints (1841, 1846), and which within three years of its first publication was translated into German, French, Italian, Spanish, Swedish and Russian. As shown by N. Rosenberg's analysis of the citations of Babbage's work in Karl Marx's *Das Kapital* (1846), not only did the information he supplied on the real conditions of English industrial production provide an important source, but above all his considerations had a strong impact on Marx's analysis of the position of the 'human factor' in the productive process, and thus on the more specific aspects of Marx's work as a social denunciation. However, from the point of view of creation of a form of industrial engineering at the time of publication of *On the economy of machinery and manufactures*, the prevailing cultural environment was not particularly favourable to the quantitative type of studies proposed in Babbage's work[24].

The same fate awaited several isolated studies carried out using mathematical techniques by engineers on the organization of transport or on industrial-location analysis. Examples of these, in the 1830s and 1840s, are contained in the work of Charlemagne Courtois referring to the choice of certain itineraries in communications paths in which it is attempted to achieve decision-making (*choix*) criteria by formulating the problem in the mathematical language of optimization, and in the thinking of Jules Dupuit (1804–1866), on measuring utility. At the end of the 19th century, a work like *Mathematische Begründung der Volkswirthschaftslehre* (1885) by the engineer Wilhelm Launhart (1832–1918), which gathers together an exposition of problems of a general nature concerning the functioning of the economy, as well as an analysis of concrete transport and production problems (with the help of isoperimetric methods), remained unknown for a long time[25]. Particularly interesting, also because of the role he played in general economic equilibrium theory, is the case of Dupuit's contribution; he started off from his experience as a state engineer, and considered that political economy had the task of examining the problem of the conditions to be satisfied by a work "of public utility", incorporating it in a vast project aimed at mathematicizing political economy on the model of mathematical physics. His work did not meet with success, as has been shown by Ingrao and Israel (1990: 73–74):

> Jules Dupuit's contribution to the new science of political economy, the daughter of mathématique sociale, provides an example of how such an approach, although persecuted and proscribed in the circles of official physico-mathematical and economic sciences, survived at least in part in some of the circles of applied science produced by the Écoles, which were sensitive to the processes of interdisciplinary interaction generated by the Revolution.

The failure to develop a mathematical theory of management that could be applied both to problems of manufacturing and of transportation is thus linked to the decline of the project of mathematization of the social sciences of the Enlightenment, which was also the main reason for the lukewarm reception of the project to develop a mathematical economics, previously formulated by Isnard, Dupuit and Augustin Cournot (1811–1877), later resumed and strenuously defended by Léon Walras. In the same year as Launhardt's book was published, Walras began correspondence with a French engineer-economist of the 'second generation', Emile Cheysson (1836–1910); Cheysson's attitude is indicative both of the persistence of a specific tradition among engineers, to which reference has already been made, and of the limits of such a tradition. Cheysson's interest in economic problems and in quantitative tools was closely related to practical needs. Former director of the statistics service at the Ministère des travaux publics, with active directive and management responsibilities in both the public and the private sectors, he was the first professor of industrial economics at the École des Mines in Paris. In July 1885 Cheysson described his approach to Walras as follows[26]:

> As our mutual friend M. Haton de la Goupillière has too kindly told you, I have tried for my part to use graphical procedures to approach a certain number of industrial problems, such as finding an advantageous rate of remuneration for a product, of the quantity of labour. In this research I have sought to eliminate purely mathematical abstractions and speculations and to implement statistics no longer in a passive state but in an active state, which in the basis of experimental data can determine laws, the progress of certain phenomena, and with the aid of intersections or inflections in graph curves – interpolated if necessary beyond their experimental area – provides solutions to problems faced daily by manufacturers.

In a letter to Haton de la Goupillière he gave an explicit description of the philosophic basis underlying this approach[27]:

> In principle, I hold little belief in the success of attempts aimed at enclosing within algebraic formulae phenomena in which human freedom is involved. Equations necessarily neglect a number of data which falsifies the conclusions at the same time giving them a dangerous appearance of rigour. It thus represents for me more an ingenious mental gymnastics than a thread guiding us through the labyrinth of social facts in which moral forces play the principal role and which elude calculation. We are dealing here with a special dynamics of which the laws are grounded in experience and not in mathematics.
>
> Nevertheless they may usefully be applied to particular questions in which only matter is involved, such as in Banks or Money [...].

The presence behind the scenes of this proposed rationalization of the human factor represented by the workers and by the early figures of "verifiers" or controllers and administrators is a fundamental aspect of the development of studies on industrial organization: vigorously present in Babbage's book, it returns at the end of the 19th century in Cheysson (who, under the influence of Frédéric Le Play, was sensitive to social problems and during his activity in the private sector concerned himself with improving the conditions of the workers), as well as in Frederick

W. Taylor, Henri Fayol and the other pioneers of modern industrial engineering. Enlightenment style reformism accompanies in a natural fashion the desire for intervention, for control, the rational nature of which is guaranteed by quantification, albeit often in the softer form of statistics rather than that of mathematical analysis: statistics seems to guarantee adherence to reality and to the empirical data, avoiding the cold abstraction and reductionist effects of mathematical equations. Nevertheless, in Cheysson's words the limits of the statistical tools available at the time are exposed: he actually speaks of the need to proceed beyond "the passive state" (which could correspond to descriptive statistics), in order to arrive at an "active state". It was not long before such techniques were developed.

The difficulties encountered in the programme to mathematize sectors of knowledge lying outside the physical sciences no doubt depended also on the mathematical tools available; even Condorcet considered that it was not possible to transfer the mechanical and physical mathematical model to questions involving the moral sciences, and that perhaps probability calculus could be of assistance in dealing with this kind of problem. It is no coincidence that one offshoot of the thinking of the Enlightenment was the development of economic, health and demographic statistics techniques, which represent an effort to introduce elements of quantification into the management of society and the economy. Starting with Adolphe Quetelet (1796–1874), contributions were made to this development by state functionaries (such as the Italian Gioia, cited by Babbage), actuaries and engineers. This kind of scholar was also a reformer and a "rationalizer" in both the public and the private sector, both in the development of national infrastructures and of public welfare, of mutual societies or insurance companies, or else of industrial plant and transport networks. Thus, it may be claimed that the project to apply mathematics to the economic and social sciences – ranging from the simplest forms of quantification to the attempts to 'write equations' – lived on in silence in the areas of practical applications and in connection with the development of such typically 19th century professions as the modern state bureaucracy, actuaries and engineers; in the meantime, the field of economic scholarship that was gradually emerging (replacing the classic *sciences morales et politiques*) was long dominated by the historical school refractory to mathematization.

The possibility of fruitful interaction among the theoreticians of economics and those concerned with the problems raised by concrete economic activities was dependent on the dual nature of the descriptive approach and the normative approach to economic and social reality. A regulatory type of economic analysis could actually quite naturally satisfy the need for control typical of the engineer's approach, now extended to include also the problem of the organization of industrial activities, public works and so on. From both these points of view (normative economics, activities engineering) mathematics was offered as a natural tool, as it represented the concrete possibility of setting action criteria (decision-making criteria, in 20th century language, or else "rational choices"). To use Cheysson's words, the curves could help solve problems with which the industrialist has to cope every day, curves that, although based on experimental data, could be "interpolated outside their experimental areas". One important obstacle nevertheless remained, namely the epistemological problem underlying the suitability of the mathematical framework for economic and social phenomena. When, in the early 20th century, this epistemological problem that had been so important in the previous century was swept away by the development of the modeling approach, mathematics seemed more than ever the action tool to be developed.

Walras, although defending the need to mathematicize economics, by means of a 'strong' mathematization programme based on the introduction of algebraic equations and not only on quantitative and statistical techniques, still hesitated between a descriptive approach and a decidedly regulatory approach. Nevertheless, perhaps also as a result of the campaign he launched in all directions to find a channel through which to spread his ideas, in his correspondence with Cheysson he mentions the link between economic theory and the "economics of manufactures") (the management of industrialization) we saw in Babbage[28]:

> I increasingly believe that there are quite a lot of these mixed enterprises in which free competition is not sufficient and that, on the other hand, the State is not a suitable manager, but it is necessary to combine individual and collective interest. It is just that it is difficult to proceed to this combination rationally, saying in each case why and in what conditions one does so. This presupposes a pure political economy and a social economics that are highly perfected.
>
> The scope of our science extends vastly from elements of equilibrium and the abstract foundations of social wealth down to the infinite details of production and the conservation of this wealth. One is entitled to search one portion or another of this huge territory in depth without however losing sight of the general view. This is what I try to do myself. I hope that, for your part, despite your obvious interest in applied questions, you will not overlook those of the principles.

2.4 Measurement and Mathematical Techniques: The Birth of Industrial Engineering

The historical parabola of the early attempts to introduce quantification into the field of engineering, management and the organization of labour and production follows the better known one of the programme of the mathematization of the economic and social sciences. In both cases we are dealing with projects put forward in the intellectual climate of the Enlightenment, that later encountered much active opposition and a general indifference and that were continued in various forms by isolated workers, and that enjoyed a great revival in the 20th century. A similar parabola may be found also in the case of another important sector of the non-physical sciences, namely the biomedical sciences, in particular, the biology of populations: also in this case, between the 18th and the 19th century we find several pioneering contributions produced by the vast project of mathematization, such as the research by Daniel Bernoulli (1700–1782) – a follower of Condorcet – on the effectiveness of inoculation against smallpox, or else that of Pierre-François Verhulst (1804–1849) – a follower of Quetelet – on the logistical growth of populations, which remained at length without any follow on. Indeed, the greatest novelties from the point of view of the renewal of the mathematical tools available that emerge between the end of the 19th and the early 20th century originated precisely in the sector of the life sciences.

Renewal of the mathematical tools was, as we have seen, a necessary condition for the continuation of a programme of mathematization of the non-physical sciences. Indeed one of the most serious obstacles to the spread of Walras' work was precisely the mathematical weakness of his approach, which was pointed out also by the mathematicians that he consulted. This need was strongly felt by those who laid the bases of modern biometric techniques, Francis Galton (1822–1911) and Karl Pearson (1857–1936), and biomathematics, Vito Volterra (1860–1940). Nevertheless, their contribution led to a fundamental split in the approach to this renewal. Volterra adopted a reductionist type of approach in which the path to be followed was that of extending the conceptual model of mechanics and of mathematical physics to the non-physical sciences; for him, it was a matter of adopting mathematical analysis and differential equations. He explicitly defended this approach as early as 1901, and it formed the basis of the great scientific project he undertook in 1925 to pave the way to a theory of population dynamics. The English biometricians, on the other hand, chose to address different types of quantitative and mathematical tools developed to handle observed scientific data or else demographic and social data. These were thus tools based essentially on probability and error theory, which still occupied a marginal position among mathematical disciplines. Pearson identified them as an alternative to the causalism and determinism involved in the mathematization of mechanics and a basis for the extension of science to all fields of knowledge. This "programme" formed the basis of his *Grammar of science* (1892) and was applied throughout his scientific life.

The theoretical work done in this period is fundamental for the later extension of mathematization. Indeed, Volterra succeeded in the field of biology because he made a fundamental contribution to the formulation of a biological theory in terms of systems of nonlinear differential equations, thus providing a self-sufficient example – regarding population dynamics – of the path to be followed in the mathematization of non-physical sciences from the classical standpoint. For their part, the English biometricians began to develop statistics in the modern sense, that is, "as an area of mathematical enquiry, standing above every particular application" (Porter 1994: 1335), as represented by the seminal work of Ronald A. Fischer (1890–1962) *Statistical methods for the research worker* (1925).

The radical difference between these two approaches may be seen also in the reactions produced in scientific circles in the 1920s and 1930s by Volterra's work, and in particular the debate that arose concerning the problem of verifying his mathematical results by means of observed or experimental data (Israel and Millán Gasca 2002). It may be said that, over and above the appreciation for the strictly mathematical value of his work – Volterra was a distinguished mathematician –, his methodological and epistemological proposal was not received with any enthusiasm; indeed his studies were developed only later, starting in the late

1950s. The debate showed that in those years the idea of a "strong" mathematization of the non-physical sciences, on the model of mechanics, persisted among a number of Continental European scholars, in France and Russia in particular. However, in the years prior to Volterra's theoretical proposal, another different programme had arisen in the wake of the climate of positivism and the late 19th century scientistic optimism. The idea of extending scientific rigour to all sectors of knowledge and practice again emerged with force: examples are the establishment of a "scientific" sociology and psychology, the spread of the idea of technology as "applied science", the birth of econometrics and even the efforts made by biometricians to consolidate the transition from "natural history" to biology by assigning scientific rigour to the theory of evolution. This idea of rigour was no longer based on the model of mechanics and on mathematization. The accent was now laid on the gathering of observed and experimental data to support theory in all disciplines, on measuring and quantifying; at that time, it indicated the predominance of the Anglo-Saxon empirical tradition over the scientific rationalism of the Enlightenment. This approach led to a renewed interest in the development of quantitative techniques to facilitate the treatment of empirical data and to allow their in-depth analysis: the new statistical techniques provided an answer to this need for adherence to reality.

This cultural context, which is reflected also in the words Cheysson addressed to Walras with reference to the world of industry and production, marked the emergence, after the long parenthesis following Babbage's work, of new attempts to hammer out a "scientific" approach (in the sense that we have seen to prevail in this period) to the problem of the organization of labour and production – the birth of industrial engineering. It marked the scientific management programme of Frederick W. Taylor (1856–1915), that is, the idea of introducing measurement and experimentation into the factory, not only to innovate the hardware by applying the developments occurring in physics and chemistry (as he himself had done to facilitate the operations of cutting metal, together with Maunsel White), but also to plan the activities. This was followed by the attempts to apply scientific physiology and psychology by Frank Gilbreth (1868–1924) and Lillian Gilbreth (1878–1972) and the studies carried out at the Hawthorne plant of Western Electric; and also the development of the first statistical techniques of quality control of manufactured products.

During the long parenthesis we have mentioned, strong progress was however made in the factories from the traditional organization still based on the shop model, which was described by Babbage for England, towards the new organization of industrial plants in the United States, in Britain and in Germany, with the increased number and improvements in machine tools, the growth of the physical structures and the number of workmen, the introduction of the principle of the interchangeability of the components. The human factor, which had justified the reluctance to use mathematical tools other than statistical techniques in this context, was increasingly present: the first work measurement studies were carried out by Taylor in the years 1880–81 when he was working as an engineer at the Midvale Steel company following a controversy with workers. Also other needs linked to directions of development forecast by Babbage came to a head, such as competition and increased plant complexity. The desire to come to terms with these needs by adopting a scientific approach not only led to measurements being performed and probabilistic techniques being introduced in sampling, but also to entire production *systems* being considered, without isolating the separate components, and to examine from within organization structures such as those involving labour supervision and management in the work of Henri Fayol (1841–1925) or the problems of production planning and scheduling in the work of Henry Gantt (1861–1919) [29].

The technological and organizational evolution of the factories thus began to be a subject of in-depth scientific consideration which includes also, although not always, quantitative techniques. Curiously enough, this thinking was accompanied by an almost missionary zeal: from 1893 on Taylor dedicated himself to disseminating his ideas, travelling all over the United States; Henri-Louis Le Chatelier (1850–1936) carried out an actual campaign to spread his ideas in France. The same happened for the new proposals put forward in the 20th century in this sector, for example by W. Edwards Deming (1900–1993). In any case, the idea of scientific management set the right conditions for the introduction of mathematical methods into this context. However, the nature of these methods was still widely open to debate. In the view of Le Chatelier, for example, the spread of Taylor's ideas was linked to the need to relate the formal training of French engineers with the actual conditions in the companies and industrial plants, and to the advisability of practical training periods in the factory (Moutet 1992); and yet, starting from a rigidly deterministic scientific epistemology, close to that of Volterra and many of his French interlocutors, he rejected the introduction of probabilistic methods. In spite of the position held by Cheysson, publications such as *Anwendungen der mathematischen Statistik auf Probleme der Massenfabrikation* (1927) by R. Becker, H. Plaut, and I. Runge, and above all *Economic control of quality of manufactured products* (1931) by Walter Shewhart, as pointed out by D. Bayart and P. Crépel (1994: 1388), represented an "upheaval [...] in relation to the way most engineers thought at the time". This wide ranging debate can be better understood in the broader context of the problem of the mathematization of the non-physical sciences. For example, Egon Pearson (1895–1980), Pearson's son and the author of fundamental contributions to the statistical study of standardisation and quality control, was highly critical of Volterra's contribution.

2.5 A Fresh Start: The Birth of Operations Research

Consequently, between the end of the 19th and the beginning of the 20th century the problem of operations analysis and management was explicitly acknowledged as being the task of the engineer which had to be thought through and not just left to rule of thumb solutions worked out in the workshop. The conviction emerged that it was necessary to work out a "scientific" approach to such problems, where the adjective "scientific" represented a very generic indication of a systematic approach based on measurement and the collection of empirical data, as well as on the possibility of "applying" scientific knowledge. It is customary to associate the development and the spread of scientific management with the birth of industrial engineering. However, this was a slow process and did not proceed univocally. Even in the early 20th century, engineers' handbooks contained both mathematical techniques and some elements of physics and chemistry, side by side with some concepts of law or economics/accounting which were certainly believed to represent basic techniques for coping with typical management problems related to human resources, organization structure and costing problems, and what could today be described as the information flows of a company, construction site or of any kind of plant.

In the first few decades of the 20th century, the organizational task of the engineer became even more complicated, and was extended into new technological areas. It was no longer only a matter of managing a mine, a construction site for a large public works project, or the development of the national rail network. In the last case cited, the problems related to the design and management of a technological or 'artificial' reality having the characteristics of a "system" are already clearly present; it is no coincidence that during the second half of the 19th century it gave rise to contributions that foreshadowed many modern issues. In the early 20th century the number of these real instances increased with the development of commercial telephony in the 1920s and 1930s and the spread of mass production. World War II raised new problems in old contexts (renewed need for flexibility that clashed with modern production line systems; shipping movements in a military context; logistics problems involving unprecedented numbers) or radically new problems (radar-assisted anti-aircraft defence; anti-aircraft fire control devices). This resulted in the emergence of new mathematical tools for systems management: development took place at a dizzy rate following several different theoretical approaches and in several different cultural and applications contexts. Control engineering, mathematical programming and operations research represent different aspects of the new "system approach".

In what sense may we speak of "a fresh beginning" in the history of the application of mathematics to management and organization problems? In the sense that, in this period, we witness the resumption of the central problem of the planning and operation of a large and complex activity defined as a scientific problem; and further impulse is given to the attempt to introduce mathematical tools, starting with the classical optimization methods. This problem and these tools had already been identified in the first examples of "operational problems" in military matters and in the early factories between the late 18th and the early 19th century. Now, however, the mathematical tools are diversified, reflecting the evolution in mathematics that occurred between the late 19th and the early 20th century with the development of modern algebra, the birth of mathematical logic and the resumption of studies on discrete mathematics. Side by side with classical optimization techniques of the analytical type there were now algebraic techniques like linear programming (maximization of a linear function subject to linear inequalities) and game theory[30], or logical combinatorial mathematics, such as network flow analysis, and lastly probabilitistic techniques, derived in part from control engineering.

The studies carried out so far point to the existence of two main cultural contexts for this fresh start – that of the Soviet Union and that of Britain and the United States. In the Soviet Union the dominant belief was in the classical application of mathematics according to the mathematical physics model, which also explains the continuity of research in mechanics, in which the control theory studies were carried out. At the same time, a highly innovative line of research existed towards a new holistic scientific paradigm gravitating around the idea of system: this forms the basis of the work by Vladimir Vernadsky (1863–1945) in the natural sciences, but also of the proposal of Alexander Bogdanov (1873–1928) for a *tektology* or science of organization[31]. As far as actual mathematical studies on the problem of organization are concerned, during the 1920s and 1930s contributions were also made by various authors to the classical problems of scheduling railway traffic and industrial production, including the outstanding work of Leonid V. Kantorovich (1912–1986), who introduced linear programming and in 1939 published his *On Mathematical Organization and Planning* (Brentjes 1985). However, in the years that followed, the application of mathematics to the non-physical sciences ran into serious opposition of an ideological nature in the communist world, which also explains the delay in embracing cybernetic ideas.

The scientific work programme formulated in the United Kingdom in the pre-war period and during the war under the name of operational research represents a turning point in mathematical applications to management and organization problems (see chapter 4). It originally arose in a cultural context with which we are familiar, linked to problems of managing military operations: while the French engineers of the Enlightenment attempted through their scientific and mathematical approach to contribute also to the solution of organization problems in the military field, during World War II in England and the United States the same conviction arose of the utility of a scientific approach to the problem of organization. A definition dating back to 1948 provides an example of this: "the application of the scientific method to finding the most economical and timely means for getting the maximum military effect from the available or potentially available resources in matériel or personnel" (see Rider 1994: 837). However, this does not consist solely of a generic faith in scientific élites nor of an extension of ideas concerning the scientific organization of production. Also here we see an explicit reference to the mathematical formulation of these problems that grew up between the 18th and 19th centuries. It was present in the reports on operational research dating back to 1941 by the British physicist Patrick Blackett (1897–1974), couched in terms referring to the classical tools (albeit combining differential equations and the variational approach with probability and statistics). It was to be developed mainly in the United States (using the slightly modified name of operations research) where new mathematical tools were later introduced, starting with the linear programming developed by George Dantzig (b. 1914), independently of Kantorovich, in order to solve on behalf of the U. S. Air Force the problem of "the construction of a schedule of actions by means of which an economy, organization, or other complex of activities may move from one defined state to another, or from a defined state toward some specifically defined objective" (Wood and Dantzig 1951: 15).

Several aspects should be emphasized in order to illustrate more clearly the reasons for this new lease of life. It was part of a general resumption of interest in the applications of mathematics to the non-physical sciences linked to the crisis in classical physical mathematics and to the development of a modeling approach in applied mathematics, which became firmly incorporated in USA scientific and engineering practice. The working method based on the construction of models allowed all epistemological preclusions to be overcome. As von Neumann wrote in the mid 1950s[32], "the sciences do not try to explain, they hardly even try to interpret, they mainly made models". The model is a mathematical scheme representing several aspects of a phenomenon, without any claim to provide global explanations, and certainly not of a causal nature; again following von Neumann, what is expected is that it will work, as well as satisfy certain "aesthetic criteria – that is, in relation to how much it describes, it must be rather simple". The idea was accepted naturally also by mathematicians as it provided a new conceptual framework within which to view the relationship between mathematics and reality; an idealistic type of framework (mathematical models as "abstract schemes of possible reality" in Bourbaki's words) that was consistent with the Bourbaki axiomatic conception of mathematics. Even if the paradigm represented by mechanics and by mathematical physics in general continued to be present in the work by many of the authors who ventured into mathematically unexplored territories, the crisis in mathematical physics was to release enormous energy, opening up the way to a bolder attitude regarding the use of mathematical tools and also to the relationship between theory and empirical verification.

Between the 1920s and 1930s, mathematical studies on general economic equilibrium theory received a strong impulse from the socalled Vienna School. A decisive contribution was made also by the publication of *Theory of games and economic behavior* (1944) by von Neumann and Oskar Morgenstern (1902–1977). At the same time, econometrics was being developed in the United States through the fundamental contribution of Wassily W. Leontief (b. 1906). Again, this time in the field of biology, the early decades of the century witnessed not only the contribution by Volterra but also those of Alfred J. Lotka (1880–1949), the author of *Elements of physical biology* (1925), and the founders of populations genetics Ronald A. Fischer, John B. S. Haldane (1892–1964) and Sewall Wright (1889– 1988). However, the modeling approach was particularly well suited to the need to obtain assigned behaviours in the systems, which was typical of contexts involving engineering work and in technical-organizational problems (see chapter 3). It is no coincidence that the progress made in operations research during the first stage of its development arose out of an accumulation of models made to measure of concrete situations. The modeling method, with its bold use of mathematical techniques, above all makes it possible to transfer models that work from one problem to another, apparently distant, within the same subject matter or even between different ones. This form of growth of operations research gives it its characteristic eclectic appearance, which also makes it difficult to define. R. Rider has spoken of the "changing contours" of this field. A first change in approach and content is suggested also in the different terms used for it in the United Kingdom and the United States.

We are actually dealing not with a new sector of investigation regarding a given group of phenomena but rather with the re-emergence of a cultural project based on the assertion of the need for mathematical rationality as a basis for action; of a project that would only gradually become filled with content in the form of models, theories, fundamental theorems and methods of investigation. At the same time, even if dictated by needs for flexibility in the use of mathematics, the modeling approach contributed to drive operations research towards an abstract point of view regardless of the concrete applications, the image of models – abstract schemes – that are treated and of the axiomatic mathematics comprising their raw material. Consequently, in the subsequent development the memory of the old project was lost. The best proof of this is provided by the historical reconstruction of the origins of operations research, as elaborated by the main figures responsible for these developments during the phase of institutionalisation of the discipline. Indeed they turned to the past not so much in order to seek the origins of the problems to which they provided an answer, but to the origins of the mathematical methods themselves.

While the modeling approach contributed to releasing the energies required to develop ad hoc mathematical techniques to solve problems of military and industrial management and programming both in the Soviet Union – during the post-revolutionary period of the country's construction and modernization – and in the postwar United States, a further impulse was provided by the interaction with economics research. This interaction was catalyzed by two factors: on the one hand, the sharing of mathematical tools; on the other, the shared interest in the problems of planning and control. This interest echoes the programme of *mathématique sociale*, together with its reformulation by 19th century engineers, that is, the proposal of mathematics as the key for a rational social choice. Indeed, Leontief's ideas are traced back to François Quesnay's *tableaux économiques*, while mathematical economics studies, particularly in the seminal work of von Neumann and Morgenstern, are set in the context of a new social science defined as a social theory of rational decisions, the heir of Condorcet's *mathématique sociale*. This research quite naturally linked up with operational problems, where it is necessary to cope, on another scale, with an organizational problem related to human and economic-accounting problems. Dantzig's intellectual itinerary is revealing in this connection: he based his first model on Leontief's input-output model of the national economy; and von Neumann, for his part, showed him the link between his model and the theory of games. The explicit acknowledgement of this link – from "the abstract foundations of social wealth down to the infinite details of production" using Walras' words – came with the linear programming conference held in Chicago in June 1949, at the Cowles Commission for Research in Economics. The book in which the contributions were published, which opens with a report on Dantzig's fundamental work, was entitled "Activity analysis of production and allocation", which simultaneously captured the dual theoretical economics and management aspect of this new line of research. At the beginning of the book, Tjalling Koopmans formulated the common theme, defining it as "a fundamental problem of normative economics: the best allocation of limited means towards desired ends"; and he emphasized the unexpected links among the work developed independently by economists, mathematicians, and administrators. Among the various lines of thought, he mentioned the Vienna School, welfare economics, Leontief's work and research on the organization of defense or the conduct of war; he wrote (Koopmans 1951: 4):

> There is, of course, no exclusive connection between defense or war and the systematic study of allocation and programming problems. It is believed that the studies assembled in this volume are of equal relevance to problems of industrial management and efficiency in production scheduling. They also throw new light on old problems of abstract economic theory. If the apparent prominence of military applications at this stage is more than a historical accident, the reasons are sociological rather than logical.

In the years that followed this type of research was to be developed in several directions in the United States. In the military sector, particularly at the RAND, for the purpose of constructing a "theory of military worth", the idea was developed of providing a mathematical basis for decision-making; the main tool was identified as game theory; and the range of research was extended to take in the mathematization of economics and the social sciences (Hounshell 1997). The organizational-management aspects played an important role in the planning of the great US defence projects; as A. Hughes and T. P. Hughes (2000) point out, this planning combined mechanical, electrical and organization aspects within the system; in the 1950s a systems approach emerged in which engineering aspects were combined with those of control and the sciences of management. In the civilian sphere, a systematic programme was launched to transfer the new mathematical tools to the field of industrial organization. The sharing of control objectives – the "close control of an organization" to use Wood and Dantzig's words (1951: 15) – together with modeling methods contributed to the transfer of mathematical models; and many control engineers, resuming a tradition that we have seen to date back to the 19th century, used their technical background to approach the problems of management and of the related theoretical economic problems Klein (2001). The interaction between military research, industrial engineering and economics in the field of organization was constant and comprehensive, and is still comparatively little known; it was favoured by the meeting of a prescriptive technological approach, the normative aspect of economics and the mathematics of optimization. The mathematical analogy that cuts across the phenomena, although purely formal, suggested conceptual links among sectors of scientific and technological knowledge that were apparently often far apart and pushed in the direction of a new conception of the unity of science (including engineering sciences) of a neo-reductionist type, which appeared in the form of new key words such as information, decision-making, system, complexity. Despite the declining enthusiasm for the systems approach in the concluding decades of the 20th century, and the attention subsequently focused on thinking in the organization field linked to problems of competitiveness (especially between the US and Japan), many of these words have entered current language and culture due to the all-pervasive nature of the modern theories of organization.

Notes

[1] The terminology varies also as a function of the differences in the way the engineering professions developed (and in the training of the engineers, itself related to the professional profiles) in France and in Continental Europe on the one hand, in Britain and the U.S.A., on the other; see Lundgreen 1990.

[2] In *Littré* a quotation is made from Voltaire's *Henriade*: "The marshal of Vauban, born in 1633, the greatest engineer who ever lived, fortified, by his new method, three hundred ancient places, and has built thirty-three".

[3] For this see Porter 1995. Porter considers the relationship between administration, technical expertise and quantification, examining the activity of the Ponts et Chaussées corps in France in the 19th century and the role of the U. S. Army engineers in the development of cost-benefit analysis.

[4] A comparative study based on considerable recent research on the topic also from the point of view of the professions is presented in Lundreen 1990. See also, in relation to the definition and the images of "technological knowledge", Seely 1993.

[5] The classical nucleus of applied mechanics (resistance of materials and theory of elasticity, statics and structures, machine theory and kinematics of mechanisms) were gradually extended by the addition of hydrodynamics, heat theory, chemistry, electricity, etc (Channell 1989). For descriptive geometry, see Sakarovitch 1998. As regards to the role played by the development of teaching in École Polytecnique of Paris in the cultural development of the engineer, see for example Belhoste, Dahan-Dalmedico, and Picon 1994.

[6] Industrial engineering is linked to management science, and sometimes represents the same thing; from an institutional and cultural point of view, the latter is more closely linked to economics and business studies, although both share a common origin in the Taylorist programme; see Urwick and Brech 1945; Doray 1979, Nelson 1980, Waring 1991.

[7] For economics, concerning the theory of general economic equilibrium, see Ingrao and Israel 1990, and for biology, concerning population dynamics, see Israel and Millán Gasca 2002.

[8] The quote is from the foreword written by John R. Crookall for the posthumous edition of the last book by John L. Burbidge (1915–1995), one of the pioneers of the development of organization technologies such as "group technology", Production Flow Analysis or Periodic Batch Control (Burbidge 1996: v).

[9] See the introduction to this volume, Biggart, Dudley, and King 1998, Levin 1999 and A. Hughes and T. P. Hughes 2000.

[10] Grattan-Guinness 1990. The expression "division of labour" had already been used previously, in particular by W. Petty (1623–1687), one of the inventors of political arithmetic, see Ingrao and Ranchetti 1996.

[11] Perronet 1739, 1740; see also Réamur 1761.

[12] Monge 1784. *Déblai* indicates the volume of soil that must be transported and *remblai* the space it will occupy after transport. This work by Monge actually dates back to 1776, and is linked to his teaching activity at the École du génie de Mézières; R. Taton concludes his analysis of this work, which is little known even though it contains an important contribution from the point of view of the theory of surfaces, writing that it represents, "sous son triple aspect, technique, analytique et géométrique un des travaux les plus originaux et les plus caractéristiques de Monge" (Taton 1951: 203).

[13] Garner 1954; the connection with the development of accounting is explored in Porter 1995.

[14] The book ran into several editions (2nd, 1832; 3rd, 1833, 4th, 1835); in quotations we refer to the 1835 edition contained in *The works of Charles Babbage* (Babbage 1989).

[15] The significance of Babbage's book for the history of economics is discussed in Rosenberg 1994; see also De Liso 1998. Babbage also devoted two books to life insurance, in 1826, and to the problem of taxation, in 1848. Rosenberg and De Liso emphasize the comparatively little known role played by the 1832 book. In the following, consideration will be given to its significance in the context of the mathematization of problems of industrial organization, the peculiar cultural approach followed by the work and the channels through which it was disseminated.

[16] *The works of Charles Babbage*, vol. 8 (Babbage 1989), p. v.

48 Ana Millán Gasca

[17] Ibid., p. 85.

[18] Ibid., p. 86.

[19] Ibid., p. 125. Babbage makes a point of emphasizing, in a note referring to Gioia, that the latter arrived independently at this principle, which plays a central role in his work.

[20] He dates the facts to the manufacturing of pins in France in about 1760 (see n. 11 above).

[21] Ibid., p. 80 ff. Babbage cite's Coulomb's article published in *Mémoires de l'Institut*, vol. II. The modernity of this chapter is emphasized by Rosenberg.

[22] Babbage refers to *Note sur la publication, proposée par le gouvernement Anglais des grandes tables logarithmiques et trigonometriques de M. de Prony* (1820)

[23] Picon 1989, in which it is emphasized that the efforts to quantify the resistance of materials and the productivity of human labour lead "à la constitution d'un nouvel espace de référence que l'on peut qualifier de proto-économique. Des préoccupations de maîtrise dynamique des coûts dépassant la pensée simplement comptable des techniciens de l'âge classique se font jour, sous l'égide de cette notion d'utilité dont se réclament tous les ingénieurs sans exception" (p. 158).

[24] Indeed Rosenberg himself emphasizes that, from the industrial organization point of view, the work did not receive the consideration it deserved, as is shown in the *Principles of economy* (1920) by Alfred Marshall; but, see, for example, Anderson and Schwenning (1936: 40–63), which shows that the ways in which this work influenced economic thinking, engineering and industrial practice still remain to be studied.

[25] See Ingrao and Israel 1990, Knobloch, Niehans, Hofmann, and Teocharis 1994, Porter 1995; Lauhardt's work was developed in *Über den Standort der Industrien* (1909) by A. Weber, translated into English in 1929.

[26] Letter from Cheysson to Walras, 19/07/1885, n. 666, in *Correspondence of Léon Walras and related papers*, see references, from now on abbreviated in CLW (my translation, also for n. 27 and 28).

[27] Letter from Cheysson to Haton de la Goupillière, 10/07/1885; n. 665, enclosure, in CLW.

[28] Letter from Walras to Cheysson, 23/04/1891, n. 1004 in CLW.

[29] Taylor 1911, Fayol 1918; Gantt 1919. The reference texts in this sector were published with some delay with respect to the actual introduction of organization technologies "on site".

[30] In fact, linear programming has a prehistory dating back also to the late 18th – early 19th c., with the outstanding contribution of Joseph B. J. Fourier (1768–1830), see Grattan-Guinness 1994b. Fourier knew of applications of what he called "the analysis of inequalities" to mechanics, the errors of observations, and elections (ibid., p. 49).

[31] Biggart, Dudley, and King 1998, Bailes 1990, Weiner 1988.

[32] See chapter 3, p. 54.

References

Anderson, E. H. and Schwenning, G. T., 1936, *The Science of Production Organization*, New York, John Wiley & Sons.

Babbage, Ch., 1989, *The Works of Charles Babbage* (M. Campbell-Kelly, ed.), vol. 8, *The Economy of Machinery and Manufactures*, New York, New York University Press.

Bayart, D. and Crépel, P., "Statistical control of manufacture", in: Grattan Guinness 1994a: vol. II, 1386-1391.

Bailes, K. E., 1990, *Science and Russian Culture in an Age of Revolutions. V. Vernadsky and his Scientific School*, Bloomington, Indiana University Press.

Belhoste, B., Dahan-Dalmedico, A., Picon, A. (eds.), 1994, *La formation polytechnicienne, deux siècles d'histoire*, Paris, Dunod.

Bélidor, B., 1729, *La science des ingénieurs dans la conduite des travaux de fortification et d'architecture civile*, Paris, C. Jombert (new edition, 1813).

Biggart, J., Dudley, P. and King, F. (eds.), 1998, *Alexander Bogdanov and the Origins of Systems Thinking in Russia*, Aldershot, Ashgate.

Brentjes, S., 1985, "Zur Herausbildung der lineare Optimierung", in: *Ökonomie und Optimierung* (Lassmann, W. and Schilar, H., eds.), Berlin, Akademie Verlag: 298–330.

Burbidge, J., 1996, *Periodic Batch Control*, Oxford, Clarendon Press.

Channell, D., 1989, *The History of Engineering Science. An Annotated Bibliography*, New York, Garland Publishing.

Correspondence of Léon Walras and Related Papers (Jaffé, W., ed.), 1965, 3 vols., Amsterdam, North-Holland Publishing Company.

Coulomb, A., 1799, "Résultats de plusieurs expériences destinées à déterminer la quantité d'action que les hommes peuvent fournir par leur travail journalier, suivant les différents manières dont ils employent leurs forces", *Mémoires de l'Institut National des sciences et arts-Sciences mathématiques et physiques*, 1e s., 2: 380–428.

Dawson, C. S., McCallum, Ch. J. Murphy, R. B., and Wolman, E., 2000, "Operations research at Bell Laboratories through the 1970s: Part I", *Operations Research*, 48: 205–215.

De Liso, N., "Babbage, Charles", in: *The Elgar Companion to Classical Economics* (Kurz, H. D. and Salvadori, N., eds.) (1998), Cheltenham-Northampton (Mass.), E. Elgar Publishers: vol. I, 24–28.

Doray, B., 1979, *Le taylorisme, une folie rationelle?*, Paris, Dunod.

Elmaghraby, S. E., 1977, *Activity Networks: Project Planning and Control by Network Models*, New York, John Wiley & Sons.

Fayol, H., 1918, *Administration industrielle et générale*, Paris, Dunod.

Gantt, H., 1919, *Organizing for Work*, New York, Harcourt, Brace and Howe.

Garner, S. P., 1954, *Evolution of Cost Accounting to 1925*, University of Alabama Press (New York, Garland Pub., 1988).

Gillmor, C. S., 1971, *Coulomb and the Evolution of Physics and Engineering in Eighteenth-Century France*, Princeton, Princeton University Press

Grattan-Guinness, I., 1990, "Work for the Hairdressers: The Production of de Prony's logarithmic and trigonometric tables", *Annals of the History of Computing*, 12 (3): 177–185.

Grattan-Guinness, I. (ed.), 1994a, *Companion Encyclopaedia of the History and Philosophy of the Mathematical Sciences*, 2 vols., London, Routledge.

Grattan-Guinness, I. 1994b, "'A new Type of question': On the prehistory of linear and non-linear programming, 1770–1940", in: *The History of Modern Mathematics*, vol. III (Knobloch, E. and Rowe, D. E., eds.), San Diego, Academic Press: 43–89

Hounshell, D., 1997, "The Cold War, RAND, and the generation of knowledge, 1946–1962", *Historical Studies on the Physical and Biological Sciences*, 27: 237–267.

Hughes, A. and Hughes, Th. P. (eds.), 2000, *Systems, Experts, and Computers: The Systems Approach in Management and Engineering, World War I and after*, Cambridge (Mass.), MIT Press.

Ingrao, B. and Israel, G., 1990, *The Invisible Hand*, Cambridge (Mass.), MIT Press.

Ingrao, B. and Ranchetti, F., 1996, *Il mercato nel pensiero economico. Storia e analisi di un'idea dall'Illuminismo alla teoria dei giochi*, Milano, Hoepli.

Israel, G., 1996, *La mathématisation du réel. Essai sur la modélisation mathématique*, Paris, Éditions du Seuil.

Israel, G. and Millàn Gasca, A., 2002, *The Biology of Numbers*, Basel, Birkhäuser.

Klein, J., 1999, *Controlling gunfires, inventories, and expectations with the exponentially weighted moving average*, Mary Baldwin College, unpublished.

Klein, J., 2001, "Post-war economics ad shotgun weddings in control engineering", Mary Baldwin College, preprint.

Knobloch, E., Niehans, J., Hofmann, A., and Teocharis, R. D., 1994, *Wilhelm Launhardts "Mathematische Begründung der Volkswirtschaftslehre". Vademecum zu einem Klassiker der Theorie der Raumwirtschaft*, Düsseldorf, Verlag Wirstchaft und Finanzen GmbH.

Kohli, M. C. 2002, "Leontief and the Bureau of Labor Statistics, 1941–1954: Developing a framework for measurement", in: *The Age of Economic Measurement* (Klein, J. and Morgan, M., eds.), Durham, Duke University Press.

Koopmans, T. C. (ed.), 1951, *Activity Analysis of Production and Allocation*, New York, Yale University Press/Wiley.

Levin, 1999, M. R. (ed.), 1999, *Cultures of Control*, Amsterdam, Harwood.

Lundgreen, P., 1990, "Engineering education in Europe and the U.S.A. (1750–1930): the rise to dominance of school culture and the engineering professions", *Annals of Science*, 47: 33–75.

MacCormick, E. J. and Sanders, M. S., 1957, *Human Engineering*, New York, McGraw-Hill (7th edition: *Human Factors in Engineering and Design* 1993).

Millán Gasca, A., 2003, "Early approaches to the management of complexity in engineering systems", in: *Determinism, Holism, and Complexity* (Benci, V. et al., eds.), New York, Kluwer/Plenum Publishers: 349–357.

Miller, D. M. and Schmidt, J. W., *Industrial Engineering and Operations Research*, New York, John Wiley & Sons, Inc., 1984.

Monge, G., 1784, "Mémoire sur la théorie des déblais et des remblais", in *Histoire de l'Académie des Sciences, Année MDCCLXXXI. Avec les Mémoires de mathématiques et de Physique pour la même année*, Paris 1784.

Moutet, A., 1992, "Rationalisation et formation des ingénieurs en France avant la seconde guerre mondiale", *Cahiers d'histoire du CNAM*, 1: 93–116.

Nelson, D., 1980, *Frederick W. Taylor and the Rise of Scientific Management*; Madison, University of Wisconsin Press.

Neumann, J. von, 1961–63, *John von Neumann: Collected Works* (Taub, a. H., ed.), 6 vols., New York, Macmillan.

Perronet, J., 1739, *Explication de la façon dont on réduit le fil de laiton à differents grosseurs dans la ville de Laigle en Normandie*, Ecole National de Ponts et Chaussées, Paris, ms 2383

Perronet, J., 1740, *Description de la façon dont on fait les épingles à Laigle, en Normandie*, Ecole National de Ponts et Chaussées, Paris, ms 2385.

Picon, A., 1987–88, "Les ingénieurs et l'ideal analytique à la fin du XVIIIe siècle", *Sciences et techniques en perspective*, 13: 70–108.

Picon, A., 1989, "Les ingénieurs et la mathématisation. L'exemple du génie civil et de la construction", *Revue d'Histoire des Sciences*, 42 (1–2): 155–172.

Picon, A., 1992, *L'invention de l'ingégneur moderne. L'Ecole des Ponts et Chaussées 1747–1851*, Paris, Presses de l'Ecole Nationale des Ponts et Chaussées.

Porter, T. M., 1994, "The English biometric tradition", in: Grattan-Guinness 1994a: vol. II, 1335–1340.

Porter, T. M., 1995, *Trust in Numbers*, Princeton (N. J.), Princeton University Press.

Réamur, R.-A. F. de, 1761, *Art de l'epinglier, par M. de Réamur, avec des additions de M. Duhamel du Monceau, et des remarques extraites des Mémoires de M. Perronet*, Paris, Saillant et Nyon.

Rider, R., 1994, "Operational research", in: Grattan Guinness, 1994: vol. I, pp. 837–842.

Rider, R., 1992, "Operations research and game theory: early connections", in: *Toward a History of Game Theory* (Weintraub, E. R., ed.), Durham (NC), Duke University Press: 225–239.

Rosenberg, N., 1994, *Exploring the Black Box. Technology, Economics, and History*, Cambridge, Cambridge University Press.

Sakarovitch, J., 1998, *Épures d'architecture. De la coupe de pierres à la géométrie descriptive. XVIe–XIXe siècles*, Basel, Birkäuser.

Seely, B., 1993, "Research, engineering, and science in American Engineering Colleges: 1900–1960", *Technology and Culture*, 34: 344–386.

Shinn, T., 1980, "From 'corps' to 'profession': the emergence and definition of industrial engineering in modern France", in: *The Organization of Science and Technology in France 1808–1914* (Fox, R. and Weisz, G., eds.), Cambridge-New York, Cambridge University Press: 183–208.

Taton, R., 1951, *L'œuvre scientifique de Monge*, Paris, Presses Universitaires de France.

Taylor, F. W., 1911, *The Principles of Scientific Management*, New York, Harper and Brothers.

Teocharis, R. D., 1994, "Die Ökonomen aus dem Ingenieurwesen und die Entwicklung von Launhardts mathematisch-ökonomischen Denken", in: Knobloch, Niehans, Hofmann, and Teocharis 1994: 55–83.

Thépot, A., 1991, *Les ingénieurs du corps des Mines au XIXme siècle, 1810–1914. Recherches sur la naissance et le développement d'une technocratie industrielle*, PhD Thesis, Paris, Université de Paris X-Nanterre;

Urwick, L. F. and Brech, E. F. L., *The Making of Scientific Management*, 3 vols., London, Management Publications Trust, 1945–1948.

Waring, S. P., 1991, *Taylorism Transformed: Scientific Management Theory since 1945*, Chapel Hill, University of Carolina Press.

Weiss, J. H., 1982, *The Making of the Technological Man. The Social Origin of French Engineering Education*, Cambridge (Mass.).

Weiner, D., 1988, *Ecology, Conservation and Cultural Revolution in Soviet Russia*, Bloomington, Indiana University Press.

Wood, M. K. and Dantzig, G., 1951, "The programming of interdependent activities: General discussion", in: Koopmans 1951: 15–18.

Zylberberg, A., 1990, *L'économie mathématique en France 1870–1914*, Paris, Economica.

3 Technological Innovation and New Mathematics: van der Pol and the Birth of Nonlinear Dynamics

Giorgio Israel

3.1 Radio Waves and Mathematical Modeling

It is sometimes claimed that the emergence of a form of mathematization of phenomena based on the use of nonlinear mathematical models resulted from (or was at least favoured by) the needs of 1940s technology, in particular during World War II. There is no doubt that the introduction of a large number of new applications or indeed of new branches of mathematics was a response to the wartime situation. The period that began in the 1940s saw the development of game theory, linear and nonlinear programming, cybernetics, the science of digital calculus, information theory, and nonlinear dynamics. It is equally well known that the notions of feedback and servomechanism played a central role in some of these developments, which thus seem to be closely related to a profound change in technological conceptions. It would however be somewhat superficial to overlook the fact that the roots of these developments originate in the earlier past, in particular in the case of nonlinear modelling and the analysis of feedback processes.

The search for the "precursors" in historical analysis leads to superficial relations being established and to the specific nature of the different contexts being concealed. The truth contained in such slogans as "Democritus is the father of modern atomic theory" is not enough to hide its sterility. In our particular case, it is not very relevant to recall to mind a few of James Clerk Maxwell's (1831–1879) intuitive thoughts to anchor the origins of the modern analysis of servomechanisms. Nor is it relevant to revisit the work of Claude Bernard (1813–1878) in order to introduce the concept of "homeostasis" (which represents the very core of the notion of feedback and thus much of nonlinear dynamics). It is true that, in both cases, we are dealing with important anticipations: however, they are linked to a conceptual and practical context that is different and remote. Nevertheless, even sticking to the "modern" form in which these ideas were explicitly expressed and disseminated during the second half of the 20th century, there seems to be no doubt that they were introduced during its early decades and not in the 1940s or 1950s. To fix the origin of the idea of homeostasis, rather than cite Norbert Wiener (1894–1964), it is necessary to revisit the work of Walter B. Cannon (1871–1945), which takes us back to the beginning of the 1930s and to topics developed in the area of human physiology. Another example is provided by the origins of such an important theory in nonlinear dynamics as "Hopf's bifurcation". Hopf's theorem (Eberhard Friedrich Hopf 1902–1983) dates to 1942, although the technological reasons underlying the theory may be sought in the study of the properties (and anomalies) of the behavior of certain servomechanisms used on steam engines at the beginning of the century. And yet it is only in the context of interest in the more advanced technologies such as radio engineering, which we will illustrate here, and of the consequent organic development of the theory of nonlinear oscillations, that this theory will be fully developed.

More generally, the theory of nonlinear oscillations and the concept of limit cycle derive from the developments in late 19th century mathematical physics, and in particular in Henri Poincaré's (1854–1912) celestial mechanics studies, as gathered together in the well-known treatise *Les méthodes nouvelles de la mécanique céleste* (1892–1899). Nevertheless, the discovery and study of limit cycles by the Dutch engineer Balthazar L. van der Pol (1889–1959), in the 1920s, took place in a completely different context and was closely linked to the emergence of new technologies. It was the study of the propagation of radio waves and of the electrical devices required to generate them that led van der Pol to work out the equation that is today considered as the prototype of the nonlinear feedback oscillator. Furthermore, van der Pol's engineering work covers such a wide area in the study of a large number of devices of great relevance to modern technology (radio, telephone, colour TV) as to highlight the continuity linking this type of research to the developments that occurred in the second half of last century.

Russian mathematical physics was strongly influenced by the theoretical work of Poincaré and Aleksandr M. Lyapunov (1857–1918) and was sensitive to their links with engineering topics: the concept of limit cycle – which emerged also in the topics related to Hopf's bifurcation – was familiar to it. Conversely, van der Pol showed no awareness – at least in the work he did in the 1920s – of Poincaré and Lyapunov's work and does not even use the term "limit cycle" to define the regime of self-oscillating electrical circuits. He uses direct methods of analysis that have no relation to what was already known in the field of the qualitative theory of ordinary differential equations. He was reproached for the "primitive" nature of the mathematical methods he used by the Soviet school, who resumed his research and subjected it to a rigorous general re-elaboration, ultimately laying the foundations for a new "nonlinear mechanics". Nevertheless, the impulse given by van der Pol's work was decisive in defining and directing the research of the Soviet school in the field of nonlinear analysis. His work proposed completely new themes and drew upon a technological context that called much more strongly and directly for the elaboration of a general theory of nonlinear oscillations (with feedback) than the context of steam engine technology. Van der Pol makes use of a rather old mathematical tool, but moves on the crest of the advanced technology wave of the time, which accompanies most of the developments occurring in the century. The Soviet researchers completely dominated the early developments of nonlinear analysis, but related them to a classical context (celestial mechanics) as well as to technological forms of the previous century. Therefore, a fundamental turning point was reached through the synergy between the new ideas expressed

in van der Pol's research and the advanced mathematical techniques developed by
Leonid I. Mandelshtam (1879–1944), Nikolai D. Papalexi, Aleksandr A. Andronov
(1901–1952), Aleksandr A. de Witt (d. 1937), Nikolai M. Krylov (1879–1955) and
Nikolai N. Bogoljubov (b. 1909) in their studies on nonlinear self-oscillating systems.

However, there is another aspect in which van der Pol's contribution seems to
be of great importance. As we have shown in other works (Israel 1996, 1998), the
views held by van der Pol on the mathematization of phenomena represent one of
the first and most precise expressions of the modern conception of mathematical
modeling. Its characteristics may be summed up in the words of John von Neumann (1903–1957): "[...] the sciences do not try to explain, they hardly even try to
interpret, they mainly make models. By a model is meant a mathematical construct
which, with the addition of certain verbal interpretations, describes observed phenomena. The justification of such a mathematical construct is solely and precisely
that it is expected to work – that is, correctly to describe phenomena from a reasonably wide area" (von Neumann 1955: 492).

Our present aim is to illustrate these two profoundly innovative aspects of the
scientific work of Balthazar van der Pol.

3.2 From Radio to Limit Cycles

Balthazar van der Pol was born on 27 January 1889 at Utrecht (Holland) where he
did his university studies, graduating in physics in 1916. That same year he moved
to England to complete his training in the radioelectricity laboratory directed by
John A. Fleming (1849–1945) at University College, London. In 1917, he moved to
the Cambridge Cavendish Laboratory, which was directed at the time by Joseph
John Thomson (1856–1940), where he was introduced to the study of radio waves.
He returned to Holland in 1919 and was awarded his Ph.D. from the University of
Utrecht for a thesis on *The effect of an ionised gas on electro-magnetic wave propagation and its application to radio, as demonstrated by glow-discharge measurement.* From 1919 to 1922 he worked as an assistant to Hendrik Antoon Lorentz
(1853–1928), at the Teyler Institute of Haarlem. In 1922 he joined the Philips
company as head physicist in the research laboratory, before becoming director of
the radio scientific research section. He stayed with Philips until 1949.

In the meantime, he also carried on academic activities as professor of theoretical electricity at the Technological University of Delft, from 1938 to 1949; in the
years 1945 and 1946 he was president of the Temporary University of Eindhoven.
In 1934 he was appointed vice-president of the Institute of Radio Engineers
(USA), of which he had been a member since 1920, and he was awarded the Medal
of Honour of this association for his contributions to the theory of electrical circuits. From 1934 to 1952 he was vice-president of the Union Radio Scientifique
Internationale (U.R.S.I.) of which he was honorary president from 1952 on, as
well as representative (1952 to 1959) on the Executive Board of the International
Council of Scientific Unions. From 1949 to 1956 he directed the Comité Con-

sultatif International des Radiocommunications (C.C.I.R.) in Geneva and was technical advisor to the International Telecommunications Union for the planning and development of radio communications in the early postwar period. After his retirement in 1956, he was invited to hold a chair for one year at the University of California in Berkeley and then at Cornell University. He died on 6 October 1959 at Wassenaar (Holland).

Van der Pol's technological and scientific interests took definitive shape during his stay at Cambridge under the guidance of Thomson. The great success achieved by Guglielmo Marconi (1874–1937) in 1901, with his radio transmission across the Atlantic, had opened up a broad field of theoretical research aimed at accounting for the nature of the phenomenon. The most important contribution was Heaviside's hypothesis concerning the existence of an ionized atmospheric layer (the "ionosphere" or "Heaviside layer") which played a role in deviating radio waves. It should be noted that the actual existence of the ionosphere was discovered only in 1925, and at that time the concept was referred to only in hypothetical terms. One of van der Pol's first research topics was precisely the investigation of the hypothesis of ionic refraction, which would allow the transmission of radio waves around the Earth. In one of his early works (van der Pol 1919–1920) he showed that, without the hypothesis of an ionized layer, there would have been serious disagreement between theory and experimental evidence. The direct proof of the existence of ionic refraction required proof of the fact that the index of refraction of the ionized medium was less than 1. Starting from the principle that air ionized by an electric discharge produces a phenomenon of refraction of radio waves, he invented an apparatus that allowed him to compute the dielectric constant of an ionized gas by means of an electric discharge and to show that it could be varied and become less than 1 (van der Pol 1919a, b). In carrying out this experiment he used a triode oscillator that enabled him to produce waves with a wavelength of about 3 metres, the shortest ever achieved so far. It is significant that this ability of van der Pol paved the way to the exploit that, in 1925, using a 200 kW triode transmitter, enabled him to establish radiotelephone contact between Holland and the Dutch East Indies, and for which he was decorated with the Order of Orange-Nassau.

What we have described above thus shows how van der Pol's research at Cambridge paved the way for the theoretical analysis of radio wave propagation in the atmosphere (through the experimental investigation of their propagation in gases) and for the theory of electron motion in triodes. This research, carried out partly in collaboration with Edward V. Appleton (1892–1965), led to a series of publications written between 1920 and 1922 mainly on the subject of oscillation hysteresis and forced vibrations in a nonlinear system (van der Pol 1920, 1922, Appleton and van der Pol 1921, 1922). Of particular importance in our case is the 1922 publication in collaboration with Appleton, as it contained an embryonic form of the equation of the triode oscillator, now referred to as "van der Pol's equation".

The authors examine a circuit of the type shown in fig. 1A and point out that, in general, the anode current of a triode is a function of both the anode and the grid potential with respect to the filament, although here there exists a fixed relation between the variable parts v_a and v_g of the anode and the grid potentials, that may

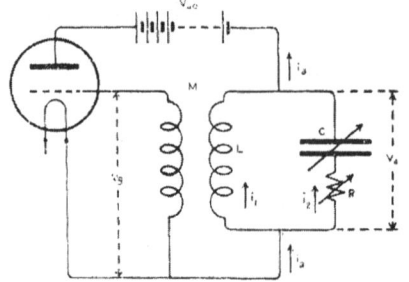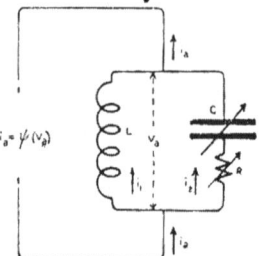

Figure 1

be expressed in the form:

$$v_g = -\frac{M}{L}v_a \, ,$$

in which it is assumed that the grid currents can be neglected. The authors observe that: "[...] in such a case the variable part (i_a) of the anode current may be expressed as a function of v_a only, and it is precisely this relation which is represented by the oscillation characteristics. [...] In this way we are able to leave out the account of the retroactive action of the control electrode, and deal simply with the problem of a conductor possessing a characteristic relation $i_a = \psi(v_a)$ connected to an oscillatory circuit as shown in fig. 1B" (Appleton and van der Pol 1922: 179).

The application of Kirchoff's law to this circuit leads to the following equations:

$$L\frac{di_1}{dt} = Ri_2 + \frac{1}{C}\int i_2 \, dt = -v_a$$

$$i_1 + i_2 = i_a = \psi(v_a)$$

from which can be derived the differential equation:

[1] $$\frac{d^2}{dt^2}(v_a + Ri_a) + \frac{d}{dt}\left\{\frac{R}{L}v_a + \frac{\psi(v_a)}{C}\right\} + \frac{v_a}{CL} = 0$$

In the case of a high frequency circuit, Ri_a is small compared with v_a and so the equation may be rewritten in the following form:

[2] $$\frac{d^2v}{dt^2} + \frac{d}{dt}\chi(v) + \omega_0^2 v = 0$$

where it is assumed that $\omega_0^2 = \frac{1}{CL}$, and $\chi(v) = \left(\frac{Rv}{L} + \frac{\psi(v)}{C}\right)$.

This represents the first form of van der Pol's equation. Analysis of the equation leads to several partial results and to the determination of the characteristic form of the variable resistance, although it does not clearly indicate the presence of the limit cycle. A decisive step forward in this direction is however contained in the paper "On 'Relaxation-Oscillations'" which he wrote on his own (van der Pol 1926). Here the author shows he now has very clear ideas concerning the nature of the process represented by the equation and about the form of the solutions.

Van der Pol considers the general form of an oscillatory system subjected to a dissipative force, namely $\ddot{x} + \alpha\dot{x} + \omega^2 x = 0$ and points out that, in the case in which the resistance is negative – as happens in certain electrical circuits (for instance, that of the triode) in which an energy input occurs – the equation becomes $\ddot{x} - (\alpha - 3\gamma x^2)\dot{x} + \omega^2 x = 0$. The solution of this second equation is however "physically unrealizable because it indicates an amplitude increasing to infinity. Thus for actual physical systems the differential equation will only be valid for values of x up to a certain value. To express the limitation of the amplitude we must assume that the coefficient of the 'resistance' term is a function of the amplitude itself becoming positive at the higher values" (ibid.: 979). In order to represent a situation of this kind, α is replaced with the expression $\alpha - 3\gamma x^2$, where γ is a constant. In this way the equation $\ddot{x} - (\alpha - 3\gamma x^2)\dot{x} + \omega^2 x = 0$ is obtained. In the case of an electrical circuit of the RLC (resistance-inductance-capacitance) type, such as a triode,

$$\alpha = \frac{R}{L} \quad \text{and} \quad \omega^2 = \frac{1}{LC}.$$

By suitably changing the unit of measure of x and of time t, the equation can be rewritten in the following form (which is the one now customarily used for the van der Pol equation):

[3] $$\ddot{x} - \varepsilon(1 - x^2)\dot{x} + x = 0 \quad \text{where } \varepsilon = \frac{\alpha}{\omega}.$$

He considers this equation as a perturbation of the oscillator with negative resistance. When x is very small, equation [3] is reduced to the form of such an oscillator, $\ddot{x} - \varepsilon\dot{x} + x = 0$; it thus presents a solution tending towards infinity and is aperiodic. When the amplitude x increases, the nonlinear term εx^2 cannot be neglected and, if $x^2 > 1$, it makes the second term of the equation positive by setting up a positive resistance and reducing the amplitude of the oscillations. He does not provide a general proof that the presence of the nonlinear term εx^2, *in any case*, regardless of the value of ε, provided that it is positive, determines a periodic solution towards which all the others tend asymptotically. He separately addresses the different numerical cases for ε, distinguishing the case in which $\varepsilon \ll 1$ from that in which $\varepsilon \gg 1$. The first case corresponds to typical triode oscillations, in which a large number of periods are required to obtain a stationary state. An approximate solution of the equation shows that it tends definitively towards a periodic stationary state. Van der Pol declares that he expects a periodic stationary case to exist also in the case in which $\varepsilon \gg 1$. Indeed, "although for small amplitudes the resistance has such a big negative value that the linear case would be highly aperiodic, the nonlinear term, i.e. $x^2\dot{x}$, makes the solution periodic. We may thus say that we are dealing with a *quasi-aperiodic* solution" (ibid.: 981). Therefore, the case in which $\varepsilon \gg 1$ can be distinguished from the others in that it is "quasi-aperiodic".

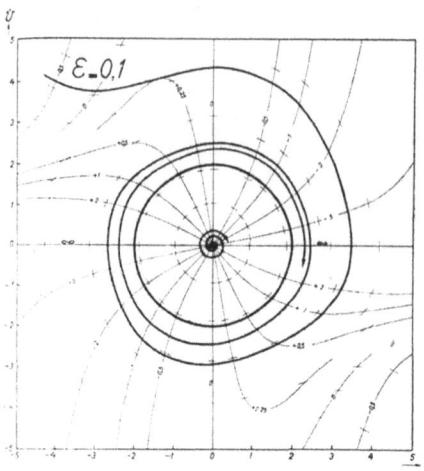

Figure 2   Figure 3

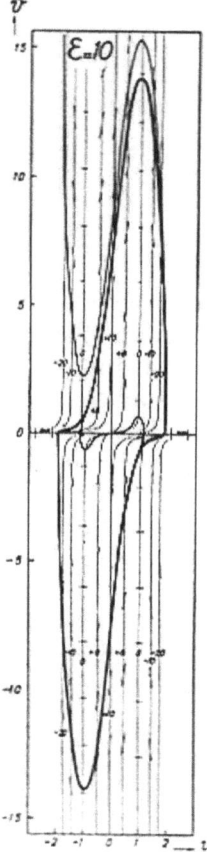

Figure 4

The explicit appearance of the limit cycle thus actually derives from the determination of what today we call the phase diagram of the equation, which is obtained using the graphical isocline method in three distinct numerical cases: $\varepsilon = 0.1$, $\varepsilon = 1$, $\varepsilon = 10$. This is perhaps the most interesting aspect of van der Pol's work, in which, even above and beyond the analytical domain of the question, there appears a complete image of the nonlinear oscillator named after him.

Let us rapidly examine these three cases and reproduce the author's original graphs. When $\varepsilon = 0.1$ (fig. 2) a quasi sinusoidal oscillation of gradually increasing amplitude is obtained, the value of which ultimately becomes defini-tively stationary at a value of 2. The case in which $\varepsilon = 1$ (fig. 3) "indicates a somewhat similar sequence of events, but here the final amplitude is reached in fewer oscilla-tions, while a marked departure from the sinusoidal form is noticed" (ibid.: 984). A particularly interesting case is the one in which $\varepsilon = 10$ (fig. 4): "Here it is noticed that the curve first rises asymptotically and after only one period practically reaches the final steady state. This steady state is characterized by a very marked departure from the sinu-soidal form" (ibid.: 987). He points out that the oscillation contains numerous higher large-amplitude harmonics and demonstrates that the period T of the oscillation, instead of being equal to 2π (as in the case in which $\varepsilon \ll 1$), is approxi-mately equal to ε and thus to RC. He calls this the *time of a relaxation*, from which derives the term relaxation-oscilla-tion suggested for this type of phenomenon.

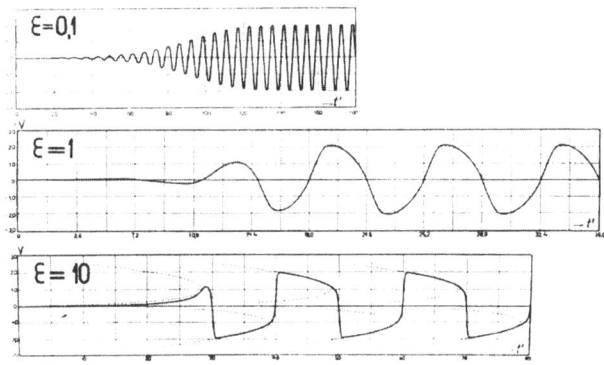

Figure 5

A conventional type representation is thus provided for the solution in all three cases (fig. 5). It should be noted how van der Pol's analysis is influenced by the specific modalities in the individual numeric cases as well as by the empirical phenomena they reflect. In the following part of the article, he actually refers to a series of devices or phenomena that could be used as examples of relaxation-oscillations. The definition could apply to Abraham-Bloch's multi-vibrator, an electric system consisting of two triodes, resistances and capacitators, that produces numerous high harmonics as it oscillates. Other examples cited towards the end of the article are "the well-known vibration of a neon-tube connected to a resistance and condenser in shunt", and "perhaps also heartbeats". This is the first reference made to the possibility of using relaxation-oscillations to model the heartbeat.

From the foregoing it is clear that this work opened up a dual line of development: on the one hand, a more detailed investigation of the general problem of nonlinear oscillations, for the purpose of working out a general theory based on rigorous and standardized mathematical methods. On the other, the extension of the field of applications of the results obtained. Van der Pol was quite aware of this second avenue of developments and at the end of his article he endeavours to provide fresh examples, in addition to that of the triode, when he suggests the heartbeat analogy. This was actually a move in the direction of modeling, in which he obtained a substantial success by virtue of the elaboration of the heartbeat model and with the suggestion of other interesting analogies. We shall deal with this issue in section 4. On the other hand, as far as the development of a general theory of nonlinear oscillations is concerned, van der Pol's contributions were affected by his limited mathematical background and, above all, by the fact that he was unaware of the contributions made by Poincaré and Lyapunov to the theory of nonlinear differential equations. It was above all the Soviet school of mathematics that took up the issues he raised and developed them in a significant way. We shall examine this aspect in a subsequent section. However, before doing so we must mention another important contribution made by van der Pol.

The contribution we are referring to is related to the van der Pol equation with a forcing term. It is generally considered to be related to work he did in 1927 and that appears to be a development of the 1926 paper described earlier. In fact, it consists of an English translation of a paper published in Dutch in 1924 (van der Pol 1927, 1924). This could explain the delay with which it became known in international scientific literature. As a matter of fact, in the 1924 paper we find the triode oscillator equation in its definitive form, albeit in the case in which it is subjected to a forcing term (namely a signal in the form of a continuous wave). To be precise, the equation is:

[4] $$\ddot{x} - (\alpha - 3\gamma x^2)\dot{x} + \omega_0^2 x = B\omega_1^2 \sin\omega_1 t$$

It is no coincidence that van der Pol mentions only *en passant* the case of zero external electromotive force – a case that will be dealt with in the later 1926 paper (itself earlier than the unmodified English translation of this article). He dwells at length on the forcing case. As M. L. Cartwright (1960) has pointed out, the method used by van der Pol to study the equation is particularly important, as it was adopted by the Soviet researchers who gave it a rigorous general re-formulation. This consisted in assuming that the form of the solution would be $x = a \sin\omega_1 t + b \cos\omega_1 t$, where a and b are functions of the time t that vary slowly. In this way differential equations were obtained for a and b, $\dot{a} = A(a,b), \dot{b} = B(a,b)$, in which the second member functions are polynomials.

This is an important article also as far as the specific results are concerned. After pointing out that, in the absence of any forced term, the free frequency oscillation ω_0 has an amplitude of a_0 obtained from the expression

$$a_0^2 = \frac{\alpha}{\frac{3}{4}\gamma}$$

he went on to say that near the resonance region ($\omega_1 \approx \omega_0$) free oscillations are suppressed by forced oscillations. Far from the resonance region, forced oscillation amplitude decreases below a_0. On continuing to desynchronize the system, the free oscillations begin to be set up with a small amplitude and a small correction of the frequency towards the frequency of forced oscillations. The upshot is that if amplitude is constant, only the forced oscillation is present. If in addition to the constant component of the amplitude a slow and periodic variation also occurs in it, then coexistence between free and forced oscillations occurs. "These oscillations – he remarks – react on one another due to the nonlinear term containing y in the equation, as opposed to cases of linear form in which this interaction is absent" (van der Pol 1927: 80).

The study of the theory of synchronization of frequency specific to a nonlinear oscillator with that of the external forcing led van der Pol describe the phenomenon of the demultiplication of frequency in the case of relaxation-oscillations. It should be noted that also this phenomenon was described by Poincaré in chapter XXVIII of the third volume of *Méthodes nouvelles de la mécanique céléste*, as Krylov and Bogoljubov pointed out in their comment to van der Pol's works. Once again, however, the issue was being tackled in a new context, above all from the technological standpoint, and it is interesting to note that van der Pol's empirical analysis succeeds in detecting the presence of a phenomenon that today goes by

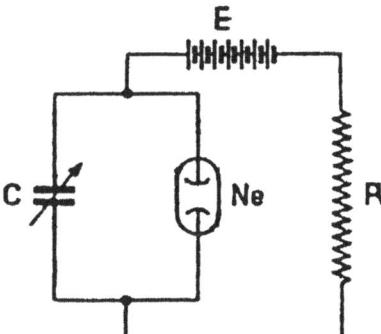

Figure 6

the name of "deterministic chaos" and the significance of which escaped not only van der Pol but even more so his contemporaries.

A paper written in collaboration with J. van der Mark describes an electric circuit which produces oscillations with relaxation (van der Pol and van der Mark 1927, fig. 6). In this case, *Ne* is a neon lamp. In the absence of an external electromotive force, the circuit oscillates with a period of $T = \alpha\, CR$, where α is a number of the order of one. If an external electromotive force is introduced, it is found that the system is capable of oscillating only at *discrete frequencies* determined by *integers sub-multiples of the external frequency*.

This is how the phenomenon is described in a later paper (van der Pol and van der Mark 1928b: 372):

> [...] we apply a small external periodic force having the same oscillation frequency as the system. By subsequently gradually reducing the period specific to the system's relaxation, we find that it continues to oscillate with the same period as the external force, so that the system automatically becomes synchronized with the external forces. The system's frequency can be decreased practically to the lower octave without preventing it from oscillating in perfect synchrony with the external force. If we continue to reduce the system's period, its frequency abruptly rises to a value that is exactly *half* the frequency of the external force and the system is automatically maintained at this new frequency over an external range. A slight reduction in the frequency specific to the system, again without modifying the frequency of the external force, abruptly makes the system oscillate on the third *subharmonic* of the applied force, where it is again maintained over an extended range, etc. [...] We have been able to prolong this experience of *demultiplying frequency* up to a ratio of 200 : 1.

In van der Pol's and van der Mark's experiment, the frequency of the oscillations was in each case determined by means of a telephone coupled to the system. They pointed out that "often an irregular noise is heard in the telephone receivers before the frequency jumps to the next lower value" (van der Pol and van der Mark 1927: 364). This observation has always escaped the attention of readers of the article. It clearly shows that they had come up against a phenomenon that may be interpreted in terms of the notion of "deterministic chaos". As is natural, the importance of the phenomenon they had detected escaped them. They actually wrote: "However, this is a subsidiary phenomenon, the main effect being the regular frequency demultiplication"(ibid.). Likewise, none of those who read the article at the time paid much attention to the observation. Even in more recent times, when J. Guckenheimer and P. Holmes (1983) dedicated their book to van der Pol, defining him "a pioneer in a chaotic land", they made no specific reference to this result.

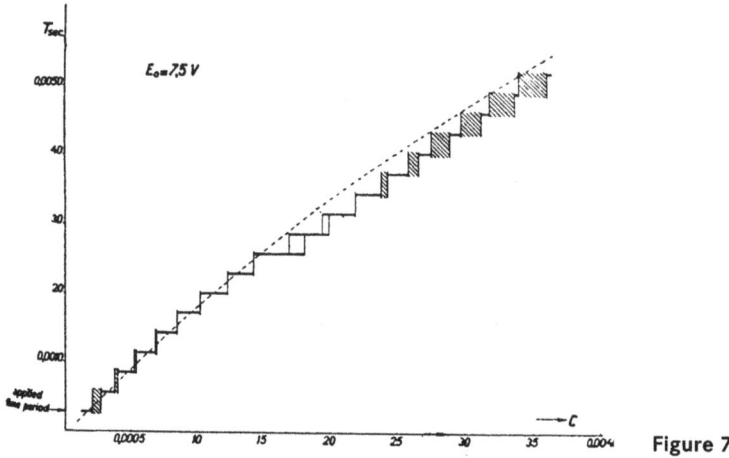

Figure 7

Fig. 7 shows a representation of the phenomenon of demultiplication of frequencies that gives rise to a discrete series of sub-harmonics[1]. The dotted line indicates the frequency with which the system oscillates in the absence of the alternating electromotive force, while the hatched portions – as the authors say – "correspond to those settings of the condenser where an irregular noise is heard" (ibid.).

In an unpublished letter written by van der Pol to the Italian mathematician Vito Volterra (1860–1940) in 1930, he speaks of the phenomenon of the demultiplication of frequencies, suggesting that Volterra took it into account in the analysis of his population dynamics models (see Appendix). There is no record of any reply from Volterra. It should be noted however that this was not the only occasion on which van der Pol showed an interest in Volterra's work on biomathematics. In a paper dedicated to providing a general outline of the nonlinear theory of electrical oscillations, he analysed the case of a triode oscillator with two degrees of freedom (van der Pol 1934). In actual fact, this analysis was a representation of the results obtained thirteen years earlier (van de Pol 1921, 1922). Now, however, he was saying that the system of linear differential equations that he had derived in order to describe the interaction between the two oscillations was "exactly equal to those occurring in a now famous problem of parasitology, where the coexistence is investigated of two species, a host population and a parasite population. This problem was investigated by A. J. Lotka and by V. Volterra" (van der Pol 1934: 1067).

It is an established fact that the Volterra-Lotka equations were derived by Alfred J. Lotka (1880–1949) and Volterra simultaneously in 1925: by the former in order to represent the dynamics between a population of parasites and one of hosts, by the latter to represent that between a population of preys and one of predators. Furthermore, Lotka had obtained the same equations in 1920 to describe the dynamics of an oscillating chemical reaction. A disagreeable controversy over priority ensued (Israel 1982, 1988). Van der Pol has the merit of not having entered the fray himself, although he would to some extent have been entitled to, having himself derived these equations in 1921. The only relevant observation that may be made is that obviously the development of nonlinear analysis was in the nature of things and that the discovery of the simplest type of nonlinear oscillator – that of Volterra-Lotka – was now to be expected in all contexts.

3.3 The Contribution of the Soviet School

As we have seen, the concept of limit cycle was already known from the work of Poincaré and Lyapunov, who had placed the problem of seeking periodic solutions for nonlinear system problems on a rigorous footing. From this point of view, van der Pol's contribution does not represent any substantial progress and indeed involves the use of non rigorous, overly *ad hoc* analytical procedures. Nevertheless, in order to appreciate its importance, it must be pointed out that Poincaré and Lyapunov's contributions had been confined to classical mechanics and mathematical physics, while the field of empirical problems, which inspired van der Pol, led to equations that were somewhat different from those usually considered. This also raised the issue of adapting and generalizing Poincaré's and Lyapunov's methods, which were considered inadequate for dealing with this new type of equation.

This is the framework within which the Soviet mathematical school was working: it did not have so much the merit of inaugurating the treatment of this new applications problem, as to have reappraised, reordered and generalized it at the mathematical level in the light of Poincaré's and Lyapunov's theory, after it too had been suitably re-elaborated. In this connection, it would seem that the peculiar value of van der Pol's contribution consisted in having ventured for the first time into a field that had hitherto attracted little attention. The Soviet mathematical school had the merit of having set the new problems in a suitable mathematical context.

One of the first articles acknowledging the relative importance of the new research was a short note by Andronov (Andronow 1929). In it the author points out that "the oscillations referred to as 'auto-entretenues' have for some years been arousing increasing interest in many fields of the natural sciences. These oscillations are governed by differential equations that differ from the ones studied in mathematical physics and classical mechanics. The systems in which these phenomena are produced are non conservative and maintain their oscillations by receiving energy from non periodic sources" (ibid.: 559). Among the phenomena to which Andronov refers, a special place is occupied by the triode oscillator, as well as periodic reactions in chemistry and biology. It should be noted that the study of these phenomena have as their point of reference the work of van der Pol, Lotka and Volterra (Israel 1993, 1988). The main purpose of the note is to show that the self-oscillations arising out of a system characterized by simultaneous different equations of the type

$$\frac{dx}{dt} = P(x,y), \ \frac{dy}{dt} = Q(x,y)$$

correspond to Poincaré's limit cycles. It goes no further, but it is intended in this way to show that "the theory of self-oscillations, in which almost exclusive use was hitherto made of non rigorous methods, is thus given, at least in the simplest case, a rigorous mathematical basis" (Andronow 1929: 561).

It should be pointed out that, in the meantime, the proof of the existence of a single period solution having the nature of a limit cycle of van der Pol's equation had been provided by A. Liénard (1928). Commenting and further developing his result, Bernard Decaux and Philippe Le Corbeiller pointed out that, by applying Poincaré's theory, it had thus been possible to obtain "a complete, mathematical and physical, explanation of the observed phenomenon" (Decaux and Le Corbeiller 1931: 725)[2].

One of the most comprehensive contributions to the problems raised in radio engineering is contained in three scientific notes by Krilov and Bogoljubov, all published 1932, and the content of which was then summarized in a subsequent article, in which the authors also presented a historical outline of the topic and made some considerations concerning the contributions made by the various national schools to the new branch of learning they called "nonlinear mechanics" (Kryloff and Bogoliuboff 1932a, b, c; 1933).

They began from the consideration that the time had now come to devote greater attention to nonlinear oscillations, which presented a number of unsolved mathematical difficulties, rather than continue to remain anchored to the now well-known theory of linear oscillations. The greatest of these difficulties consisted in the fact that in nonlinear radio circuits oscillations may be produced in which the frequencies are linear combinations of the principal frequencies, so that the mathematical functions representing them have a quasi-periodic nature. Here we see the limits of Poincaré's classic methods. Indeed the development of quasi-periodic functions following the exponents of the parameters does not provide an adequate representation and above all does not converges uniformly on the whole real axis. The original pathway indicated by the authors consists in seeking the development of the amplitudes, phases and frequencies directly rather than that of the function. The case they treat is precisely that of the van der Pol equation, for which they re-obtain in a rigorous way a series of results proved by van der Pol. The general nature of the approach nevertheless leads them to conclude that "the method by which we arrived at the results contained in these Notes are applicable and effective, as we were able to verify in the course of our research, in many other questions (for example, oscillations of synchronous machines, longitudinal stability of aircraft, etc.) and Chapters of modern physical mathematics (Quantum Mechanics), and can open the way, we think, to the creation of nonlinear general mechanics" (Kryloff and Bogoliuboff 1932c: 1122).

The foundation of this branch was presented as an achieved objective in the review article by the two authors published in 1933. After extensively underlining the fundamental importance of Poincaré's and Lyapunov's work in the direction of working out rigorous methods for finding periodic solutions to nonlinear differential equations, they acknowledged van der Pol as having the merit of being the first "to draw the attention of the world of science to the need to develop special methods for treating nonlinear problems in radio engineering, expressing himself in one of his papers as follows: 'It is therefore somewhat surprising that up to the present, though several theoretical contributions to the problem have already appeared, the phenomenon has, as far as we are aware, only been dealt with in a linear theory'" (Kryloff and Bogoliuboff 1933: 10). The limitation of van der Pol's work nevertheless lies in the fact that he did not make use of Poincaré-Lyapunov's methods, but rather of "ingenious procedures, nevertheless lacking in the necessary mathematical rigour" (ibid.). However, they also admitted that even the application of classical methods would not have been sufficient: "It is only right to observe that the comparatively non rigorous procedures of the distinguished Dutch scientist and applied by him *ad hoc* nevertheless give several indications concerning the nature of quasi-periodic oscillations, and Poincaré-Lyapunov's methods, in their present state, in no way seem applicable to the study of these objects" (ibid.).

Credit for having drawn attention to the need to deal with radio engineering problems with the required mathematical rigour was given to Liénard and Élie Cartan (1869–1951) in France and to the school of Mandelshtam and Papalexi in the Soviet Union, and mention was made of the results they obtained, as well those by Andronov and Witt. It was nevertheless pointed out that the problem remained of suitably re-elaborating the Poincaré-Lyapunov theory in such a way as to adapt it to the treatment of quasi-periodic solutions of nonlinear differential equations with self-oscillations. The results obtained in this direction seemed so solid as to justify the talk of a new branch of research to be denoted as "nonlinear mechanics" and that contains a general study of nonlinear oscillations regardless of the area in which they appear: astronomy, radio engineering, aerodynamics, chemical dynamics, animal population dynamics.

It should be noted that the terminology adopted had an obviously traditionalistic air about it. It might be claimed that it was too early at the time to use terms such as "models" or "modeling". In any case, reference to the term "mechanics" tended to include the new applications under a general heading in the classical mechanics field, rather than define a new area characterized by reference to a complex of mathematical structures. For the Soviet researchers, the emerging effectiveness of the new methods of nonlinear analysis does not seem to have led to the introduction of a new relationship between mathematics and empirical phenomena, but rather to the extension of the domain occupied by the classical mechanics approach.

Also in this respect the "less rigorous" van der Pol intervenes by playing an important role of stimulus and innovation. The way in which he proposes the use of self-oscillations (and, in particular, relaxation oscillations) in studying a vast range of "new" phenomena is also related to the idea of the mathematical model and to the practice of modelling, without any compulsory references to the schemata of mechanics, and thus taking a significant step outside mechanistic reductionism.

3.4 The Heartbeat "Model"

In an article they wrote together in 1928, van der Pol and van der Mark drew up a list of phenomena that, in their opinion, could be gathered together under the heading of relaxation oscillations (van der Pol and van der Mark 1928a: 765-6):

> Some instances of typical relaxation oscillations are: the æolian harp, a pneumatic hammer, the scratching noise of a knife on a plate, the waving of a flag in the wind, the humming noise sometimes made by a water-tap, the squeaking of a door, the multivibrator of Abraham and Bloch, the tetrode multivibrator, the periodic sparks obtained from a Wimshurst machine, the Wehnelt interruptor, the intermittent discharge of a condenser through a neon tube, the periodic re-occurrence of epidemics and of economic crises, the periodic density of an even number of species of animals living together, and the one species serving as a food for the other, the sleeping of flowers, the periodic reoccurrence of showers behind a depression, the shivering from cold, menstruation, and, finally, the beating of the heart. In all these examples the frequency of these periodic phenomena is not determined by the product of an elasticity and a mass but by some form of relaxation time.

This list is a surprising one for many reasons. In the first place because of the heterogeneous nature of the phenomena cited, which all refer to completely different contexts. These phenomena are linked together only by an *analogy* consisting in the fact that they can all be described by a single mathematical *model*. The second reason is related to the vaguely intuitive and analogical nature of the reasons why the phenomena in question may be considered as self-oscillating. On the list of "examples of typical relaxation oscillations" only a few (electrical circuit, neon tube, and the other electrical equipment) may be considered self-oscillatory phenomena on solid theoretical and experimental grounds. In all other cases, the existence of relaxation self-oscillation and a limit cycle rests solely on a vague, non rigorous intuition. The idea of representing non physical natural phenomena described on the list by means of a "van der Pol oscillator" was born and died in this article, with one single exception: the heartbeat. In this article we actually find the description of a mathematical model that became the famous prototype of a long series of mathematical models of the heartbeat which proved valid and to be of effective practical utility. The other proposals by van der Pol practically all remained a dead letter. When he speaks of the possibility of describing in mathematical terms coexistence among animal species he mentions Volterra's results. But also this reference is somewhat generic (as was that of the Soviet scientists), because Volterra's classical models have no limit cycle. The idea of describing the dynamics of economic crises by means of self-oscillations has had a rather modest outcome.

And yet, precisely this open-mindedness in seeking analogies among different phenomena, trying in this way to bring them all together within the framework of a single mathematical representation is one of the first very clear manifestations of the modeling approach. For this reason, we consider this passage to be of considerable historical interest as it expresses a clear-cut distancing from the conventional paradigms of classical reductionism (Israel 1996).

It has been said that it is only in the case of the heartbeat that van der Pol's analogy proved truly felicitous. The way this model was constructed is quite surprising and it is interesting to follow a brief description of it.

For a long time it was thought that the heart's beating was regulated via the central nervous system. However, the development of modern physiology showed that this was not at all the case: although the nervous system can affect the cardiac rhythm it plays no actual 'pace-making' role in this process. Although the heart possesses the property of *irritability* which makes it sensitive to stimuli from the nervous system, it is at the origin of its own *contractility*. Furthermore, each part of the heart is capable of generating this contraction autonomously. The different parts of the heart that have the function of stimulating contraction are nevertheless arranged in a precise hierarchical scheme. The role of "pace-maker" is played by a system of cells known as the *sinus node* (S in fig. 8). The beating of a healthy heart is regulated by the stimulus from the sinus node and is propagated through its various parts, starting from the atria (A). If the sinus node stimulus fails to reach the atria, they contract with a rhythm of their own which is slower than that of the sinus node. The same thing happens in the other parts of the heart. The sinus node stimulus is transmitted from the atria to the ventricles (V) through a bundle of muscle fibres known as the *atrioventricular bundle or bundle of His* (H). In this way, you get a rough outline of a hierarchic breakdown of the heart into five parts: more detailed breakdown may lead to more precise descriptions. The hierarchy adopted by van der Pol is the simplest possible. It is based on the hypothesis that the contractions of the atria and the ventricles are synchronous (which is only very approximately true) and schematically represents the heart as a system with three degrees of freedom: the sinus node (S), the atrium (A), the ventricle (V), with the hierarchy $S \to A \to (H) \to V$ (in which the passage $A \to V$ is not a direct one but takes place through the mediation of the bundle of His H).

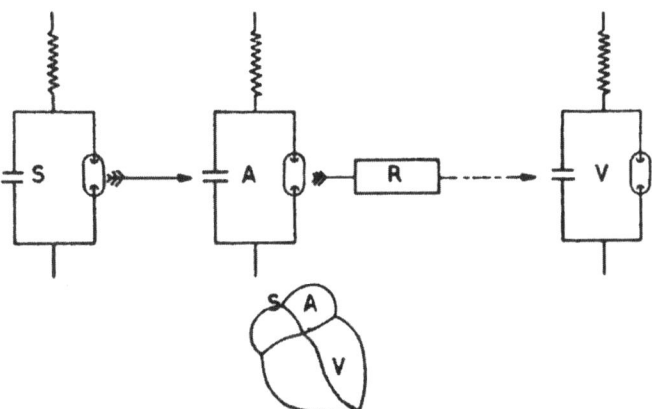

Figure 8

His idea was thus to liken the heart to an electrical system with three degrees of freedom – the three components S, A, V – each functioning like an electrical system producing relaxation oscillations. The electrical system capable of producing relaxation oscillations, and chosen by van der Pol as his model, is a circuit system of the type represented in fig. 6. Each circuit is made up of a neon lamp Ne, a condenser C with a capacitance of about 1 microfarad, a resistance R of 1 megaohm and a battery E of 180 volts. Since the condenser charging time in seconds is given by the product of the capacitance C (in farads) and the resistance R (in ohms), the period T of the relaxation oscillation is $T_{rel} \approx CR = 10^{-6} \cdot 10^6 = 1$ s. The neon lamp will thus produce one flash approximately every second.

Van der Pol's electrical heart model is thus constructed by hierarchically coupling three of these systems: the first of these represents the sinus node, the second the atrium and the third the ventricle (fig. 8). Transmission of the stimulus from A to V is represented by means of a retarding system achieved by means of a fourth circuit R containing a neon tube and, in van der Pol's words, "imitating the finite time taken for a stimulus to be transmitted from the atrium through the atrio-ventricular bundle to the ventricle" (van der Pol and van der Mark 1928a: 768).

The general electrical diagram of the model is shown in fig. 9. Each flash of the system corresponds to the activity of one part of the "heart". The system S produces relaxation oscillations. By means of the first triode, this stimulus is unidirectionally transmitted to the second relaxation system A (the atrium). The stimulation of the atrium is unidirectionally transmitted from the second triode to the retarding system R and from here to the third triode, to the ventricle V. The instrument was equipped with three keys by means of which a short electrical impulse could be transmitted to the three systems S, A, V, so as to simulate stimuli starting from different points from the usual ones and thus capable of producing an extra systole of the sinus node, the atrium or the ventricle. The R system is used to simulate lesions to the bundle of His, the socalled "atrio-ventricular blocks" ("the coupling between A and V (the auricle and ventricle) can be varied at will, thus imitating the beautiful experiments of Erlanger of gradually clamping the bundle of His" – were the observations made by van der Pol and van der Mark).

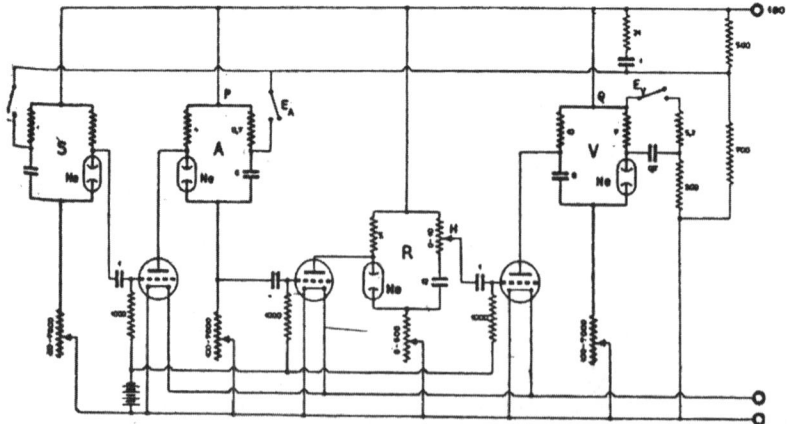

Figure 9

As can be seen in fig. 9, each relaxation system consists of a variable resistance coupled in series with a neon tube and a fixed condenser. A discharge from the latter through the neon tube represents a contraction by one of the parts of the heart. The frequency of each of the systems is regulated by the variable resistance. Two adjacent relaxation systems are coupled by means of a triode in the following way. Let us consider the sinus node-atrium coupling. One of the extremities is connected to the triode grid by means of a condenser and a grid resistance. The plate current of this triode must pass through a resistance situated above the neon tube of the atrium. In the instant in which the neon tube is illuminated, a current passes through its resistance. The potential at the lower extremity thus becomes more negative, which causes a reduction in the potential of the grid connected to this extremity and thus a reduction in the current plate of the triode. The potential difference at the extremities of its anodic resistance thus decreases, which leads to an increase in the potential difference at the extremities of the neon tube of the atrium. If at this instant the potential difference at the extremities of the tube is not too far from its discharge potential, the tube is illuminated. We then see the two tubes coming on together. If the potential difference is not high enough, seeing that the atrium condenser is still not sufficiently charged, the discharge potential is not attained and nothing happens, since the excitation occurs during the "refractory" period. We thus have an atrioventricular block.

The retarding system is also a relaxation system, although the current in it is too strong to produce oscillations. A continuous current passes through the neon tube and the system is overloaded, so that no periodic phenomenon can occur. Nevertheless, if the potential difference at the extremities of the neon tube is reduced, the luminous discharge is interrupted, but the condenser is recharged and when the discharge potential is attained the tube again produces a flash. The time elapsing between one extinction and the next discharge is the transmission delay and is equal to the relaxation time RC. External excitation thus produces only one beat. The retardation system is coupled to the ventricle in the manner described above. Since the inserted resistance is variable, ventricle excitation may be varied, as though it were constraining the bundle of His.

In fig. 9 we see that each relaxation system is provided with another resistance, like the one that, in the sinus node system, connects the condenser to the upper extremity of the high tension source. A key on the back of the apparatus allows to discharge a condenser through this resistance and to generate a stimulus in the relaxation system. Other keys can be used to generate other types of additional stimuli (or "extra systoles").

We shall skip describing the way "electrocardiograms" are made (i.e. current trend plots) by means of a special filter circuit and recorded with an oscillograph, as they are merely electrical devices that add nothing to the conceptual basis of the model.

We shall rather have a direct look at what we obtain by making these measurements, that is, the type of electrocardiograms obtained with the artificial heart and the comparisons it is possible to make with the behaviour of the human heart. The result is represented in fig. 10. The analogy with the electrocardiogram of the human heart is quite apparent. We find the main components of the latter (the P

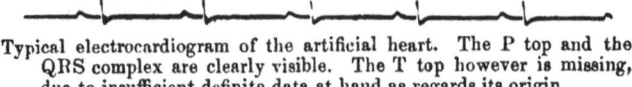

Typical electrocardiogram of the artificial heart. The P top and the
QRS complex are clearly visible. The T top however is missing,
due to insufficient definite data at hand as regards its origin.

Figure 10

wave and the *QRS* complex) with the exception of the *T* wave. In any case, the significance of the latter wave was not fully clear at the time: "as the origin of the *T* top in the electrocardiogram of the human heart is not quite certain yet, we could not insert a representing mechanism for it" (ibid.: 769).

In their work, van der Pol and van der Mark demonstrated that, by transmitting impulses to the electric heart through the keys located at the back of the apparatus, it was possible to simulate the three standard types of human heart extra systole: ventricular (fig. 11), atrial (fig. 12) and sinus extra systoles (fig. 13). They also managed to simulate lesions to the bundle of His, which in the human heart entrained the loss of part or all of the stimuli coming from the sinus and, as a result, of the ventricular beats. This loss may be an occasional event, although severe damage can lead to a systematic loss of ventricular beats. The ventricular beats may then occur at the rate of one beat every two beats of the sinus or one beat every three beats of the sinus, and so on, up to a state of complete dissociation. In the latter case, the ventricle beats at its own rate ("idioventricular rhythm"). This kind of

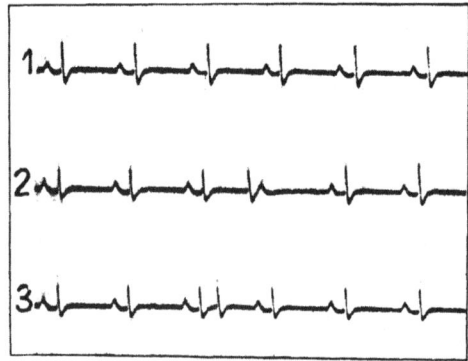

Ventricular extrasystolæ :—1, normal heart beat ; 2, late ventricular extrasystole resulting in the ventricle being in the refractory state when the
next following normal stimulus arrives from the auricle ; 3, early
ventricular extrasystole; here the ventricle is *not* any more in the
refractory period when the next following normal stimulus arrives from
the auricle and thus an *interpolated ventricular* systole is obtained.

Figure 11

situation has long been known under the name of *atrioventricular blocks*; in particular, in cases of a well-defined ratio between beats, they are known as *type n:1 Wenckebach blocks* (where *n* is the number of sinus beats which may vary from 2 to 4). Van der Pol's model allows a complete classification of these blocks to be obtained. Fig. 14 shows the results obtained. This classification was not completely known at the time of publication of van der Pol's and van der Mark's article. The model could thus be used only to predict and classify cardiac disorders that were only partially understood at the time.

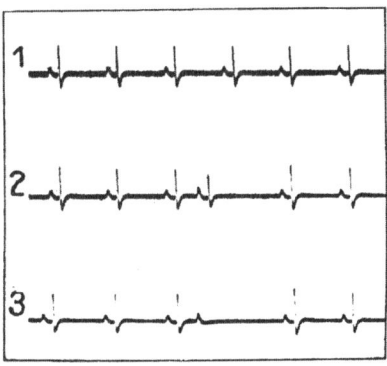

Figure 12

1. Normal heart beat.
2. Auricular extrasystole (with the ventricle responding.
3. Auricular extrasystole (ventricle still in refractory period).

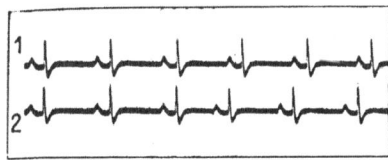

Figure 13

1. Normal heart beat.
2. Sinus extrasystole disturbing the whole heart rhythm.

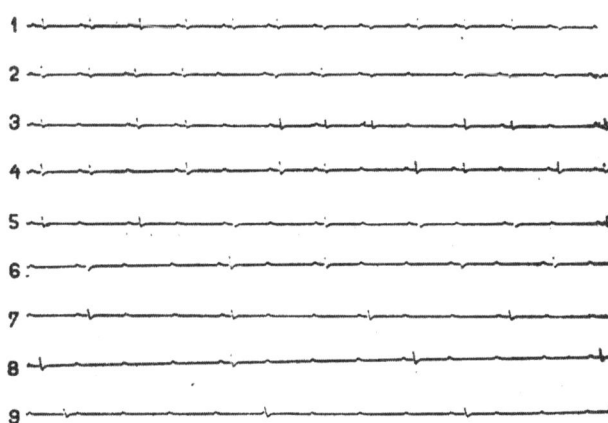

Figure 14

Electrocardiograms from the artificial heart obtained by gradually reducing the coupling between the A and V system (clamping the bundle of His. The development of 2:1, 3:1, and 4:1, as well as complete heart block, is clearly shown.

It should be noted that the heartbeat model cannot be considered a "pure" mathematical model in the modern sense of the term. Van der Pol's equation is not "the" model of the heartbeat, but only the fundamental component of its interpretation. Modelling the heart entails the (also material) construction of a physical simulation apparatus or electrical circuit. Nevertheless, the equation of this circuit is not provided, and would have to be obtained from a system of interdependent van der Pol equations. It would be a system of such complexity as to defy not only the analytical resources deployed by van der Pol, but even those available today. In any case, even via the experimental physical mediation of apparatus designed to compare the theoretical model with reality, van der Pol's heartbeat model may certainly be considered as the prototype of a form of mathematical modeling which displays two of its characteristic features: the method of *mathematical analogy,* and the special role assigned to *nonlinearity.*

3.5 Concluding Remarks

The majority of the historical reconstructions of the developments in nonlinear analysis during the 20th century followed the following stereotype. Towards the end of the previous century, important progress had been accomplished in this field by Poincaré, and later Lyapunov. These results had very little impact and practically fell into oblivion, particularly in western science, with the exception of a few isolated developments, mainly by George Birkhoff (1884–1944). On the other hand, they were taken up energetically by Soviet mathematical physics which, starting in the 1930s, also as a result of the stimulation from a number of innovative technological applications, made an outstanding contribution to nonlinear analysis. This Soviet contribution was fully acknowledged by western science only after World War II.

Although such a reconstruction contains several correct elements, a number of substantial corrections are needed.

There is no questioning the extremely important contribution Poincaré made to nonlinear analysis towards the end of the 19th century and in particular to the problem of finding periodic solutions for nonlinear systems. It must however be borne in mind that the context in which this research was carried out was that of classical mechanics. Indeed, it is incorporated in a programme for the development of analysis and in a conception of the relationship between analysis, physics and geometry that may well be described as "conservative" or at least concerned with defending the strongholds of mathematical-physical reductionism (Israel and Menghini 1998). Nevertheless, at the beginning of the 20th century, the attention of the scientific world is directed elsewhere. On the one hand, we have the developments in new physics: the theory of relativity and quantum mechanics. On the other, many scientists, particularly in the field of applied science and engineering, are exploring the implications of the new technologies: among these, radio engineering played a major role and the example of van der Pol is typical in this respect. To this situation must be added the impetuous development of mathematization in the field of the economic and biological sciences which is opening up hitherto undreamt of prospects and topics (Israel 2000). It is thus not surprising that interest waned in classical mechanics topics and that Poincaré's work in this field aroused less attention than it deserved, even to the extent of obscuring the value it had precisely in the development of that nonlinear analysis which was of such great importance in the study of the new problems.

However, it would not be correct to consider this state of affairs as a regression in western science and to attribute to Soviet science all the merit of playing a propulsive role in taking up and continuing the work of Poincaré. It should indeed be noted that the interest shown by the Soviet school in Poincaré's work was precisely a consequence of his attachment to the classical topics of mechanics and his substantial remoteness from many of the new developments in physics. It would be necessary to make a careful study of the role played by radio engineering research also in the Soviet Union in the development of nonlinear analysis, although there is evidence to believe that it did not reach the high level it reached in the West, in particular in an environment such as the Philips company. Moreover, we have seen that the Soviet scientists themselves acknowledged the decisive importance of van der Pol's work, despite the fact that the mathematical methods he used were not particularly rigorous. A necessary step was to acknowledge that Poincaré's methods, as formulated in the context of celestial mechanics, were inadequate as a basis for tackling the study of the mathematical structures proposed by radio engineering. This in no way diminishes the importance of the contribution made by the Soviet scientists; they put research on nonlinear mathematics on the right track. What we mean is that, without taking into account the powerful impulse given by the applied research developed in the early decades of the century – however *ad hoc*, non rigorous and even rough and ready – it is impossible to understand why, starting from the 1930s, there were such massive developments in nonlinear analysis and even to understand why Poincaré's work was resumed and had such a great success.

New technology and new practical applications that stimulate the development of a new mathematics: this is the theme emerging from the history of the first few decades of the century and which challenges the view of a mathematics that is always in advance of the analysis of phenomena, or, as Bourbaki (1948) put it, a set of empty forms ready to be filled out with "possible content".

Notes

1 From van der Pol and van der Mark 1928b; in 1927 a very similar but less clear one appears.

2 See also Le Corbeiller 1932.

3 It may be of interest to compare this passage with the similar one contained in van der Pol, van der Mark 1928b. Here, as well as the addition of other examples, there is another, clearer description of the analogies concerned. He thus remarks: "Let us take, for example, the aeolian harp, which consists of a taut cord against which the wind plays. A detailed examination shows that behind the cord vortexes are formed to the right and the left which are propagated away from the harp, thus making room for new vortexes that are created and that move away in the same way. Thus in the case of an aeolian harp, as in that of the wind blowing across telegraphic wires to produce a whistling sound, so the period of the sound produced is determined by a relaxation time and has nothing to do with the natural period of the cord which vibrates in a sinusoidal manner". We have here a very clear description of the socalled Bénard-von Kármán phenomenon.

References

Andronow, A., 1929, "Les cycles limites de Poincaré et la théorie des oscillations entretenues", *Comptes Rendus de l'Académie des Sciences de Paris*, 189: 559–561.

Andronow, A. and Witt, A., 1930a, "Sur la théorie mathématique des auto-oscillations", *Comptes Rendus de l'Académie des Sciences de Paris*, 190 (1): 256–258.

Andronow, A. and Witt, A., 1930b, "Zur Theorie des Mitnehmens von van der Pol", *Archiv für Elektrotechnik*, 24: 731.

Anonimous, 1959, "In memoriam Prof. Dr. B. van der Pol", *Nieuw Archief Voor Wiskunde*, (I3) 7.

Appleton, E. V. and van der Pol, B. L., 1921, "On the form of free triode vibrations", *The London Edinburgh and Dublin Philosophical Magazine and Journal of Science*, 6th s., 42: 201-221.

Appleton, E. V. and van der Pol, B. L., 1922, "On a type of oscillation-hysteresis in a simple triode generator", *The London Edinburgh and Dublin Philosophical Magazine and Journal of Science*, 6th s., 43: 177-193.

Bernard, C., 1878, *Leçons sur les phénomènes de la vie communs aux animaux et aux végétaux*, Paris, Baillière.

Bourbaki, N., 1948, "L'architecture des mathématiques: la mathématique ou les mathématiques?", in: *Les grands courants de la pensée mathématique* (Le Lionnais, F., ed.), Paris, Cahiers du Sud: 35-47.

Bremmer, H., 1960-61, "The scientific work of Balthasar van der Pol", *Philips Technical Review*, 22: 36-52.

Cannon, W. B., 1932, *The Wisdom of the Body*, New York, W. W. Norton Co.

Cartwright, M. L., 1960, "Balthazar van der Pol", *The Journal of the London Mathematical Society*, 35: 367-376.

De Claris, N., 1960, "Prof. Dr. Balthasar van der Pol: In memoriam", *IRE Transactions* CT-7: 360-361.

Decaux, B. and Le Corbeiller, P., 1931, "Sur un système éléctrique auto-entretenu utilisant un tube à néon", *Comptes Rendus de l'Académie des Sciences de Paris*, 193 (Juillet-Décembre): 723-725.

Diner, S., 1992, "Les voies du chaos déterministe dans l'École Russe", in: *Chaos et Déterminisme* (Chabert, J. L., Chemla, K., and Dahan Dalmedico, A., eds.), Paris, Éditions du Seuil: 331-368.

Guckenheimer, J. and Holmes, P., 1983, *Nonlinear Oscillation, Dynamical Systems, and Bifurcations of Vector Fields*, New York-Berlin, Springer.

Hassard, D., Kazarinoff, N. D., and Wan, Y. H., 1981, *Theory and Applications of Hopf Bifurcations*, London Mathematical Society Lecture Series, 41, Cambridge, Cambridge University Press.

Hirsch, M. W. and Smale, S., 1974, *Differential Equations, Dynamical Systems and Linear Algebra*, New York, Academic Press.

Hopf, E., 1942, "Abzweigung einer periodischen Lösung von einer stationären Lösung eines differential-System", *Berichte Math.-Phys. Kl. Sächs. Acad. Wiss. Leipzig*, 94: 1-22.

Israel, G., 1982, "Le equazioni di Volterra e Lotka: una questione di priorità", in: *Atti del Convegno su "La Storia delle Matematiche in Italia", Cagliari, 29-30 Settembre, 1° Ottobre 1982* (Montaldo, O. and Grugnetti, L., eds.), Cagliari, Università di Cagliari, Istituti di Mathematics della Facoltà di Scienze e Ingegneria: 495-502.

Israel, G., 1988, "The contribution of Volterra and Lotka to the development of modern biomathematics", *History and Philosophy of the Life Sciences*, 10: 37-49.

Israel, G., 1993, "The emergence of biomathematics and the case of population dynamics; A revival of mechanical reductionism and darwinism", *Science in Context*, 6: 469–509.

Israel, G., 1994, "Mathematical Biology", in: *Companion Encyclopedia of the History and Philosophy of the Mathematical Sciences* (I. Grattan-Guinness, ed.), London-New York, Routledge: vol. 2, 1275–1280.

Israel, G., 1996, *La mathématisation du réel. Essai sur la modélisation mathématique*, Paris, Éditions du Seuil.

Israel, G., 1998, "Balthasar van der Pol e il primo modello del battito cardiaco", in: *Modelli matematici nelle scienze biologiche* (Freguglia, P., ed.), Biblioteca Chelliana, Grosseto, QuattroVenti: 133–162.

Israel, G., 2000, "Modellistica matematica", *Appendice 2000 della Enciclopedia Italiana*, Roma, Istituto della Enciclopedia Italiana: vol. II, 196–201.

Israel, G. and Menghini, M., 1998, "The "essential tension" at work in qualitative analysis: a case study of the opposite points of view of Poincaré and Enriques on the relationships between analysis and geometry", *Historia Mathematica*, 25: 379–411.

Kryloff, N. and Bogoliuboff, N., 1932a, "Quelques exemples d'oscillations non linéaires", *Comptes Rendus de l'Académie des Sciences de Paris*, 194 (Janvier–Juin): 957–960.

Kryloff, N. and Bogoliuboff, N., 1932b, "Sur le phénomène de l'entraînement en radiotechnique", *Comptes Rendus de l'Académie des Sciences de Paris*, 194 (Janvier–Juin): 1064–1066.

Kryloff, N. and Bogoliuboff, N., 1932c, "Les phénomènes de démultiplication de fréquence en radiotechnique", *Comptes Rendus de l'Académie des Sciences de Paris*, 194 (Janvier–Juin): 1119–1122.

Kryloff, N. and Bogoliuboff, N., 1933, "Problèmes fondamentaux de la mécanique non linéaire", *Revue Générale des Sciences Pures et Appliquées*, 44 (1): 9–19.

Le Corbeiller, P., 1932, "Sur l'entretien en oscillations du réseau passif le plus général", *Comptes Rendus de l'Académie des Sciences de Paris*, 194 (Janvier–Juin): 1564–1566.

Liénard, A., 1928, "Étude des oscillations entretenues", *Revue Générale de l'Électricité*, 23: 901–946.

Lyapunoff, A., 1907, "Problème général de la stabilité du mouvement", *Annales de la Faculté des Sciences de Toulouse*, 9: 209–.

Mandelshtam, L. and Papalexi, N., 1934, "Über nicht stationäre Vorgänge bei Resonanzerscheinungen zweiter Art", *J. Zeit. für Tech. Phys.*, 4: 30.

Mark, J. (van der) and Pol, B. L. (van der), 1934, "The Production of Sinusoidal Oscillations with a Time Period Determined by a Relaxation Time", *Physica*, 1: 437–448.

Neumann, J. (von), 1955, "Method in the Physical Sciences", in: *The Unity of Knowledge* (Leary, L. ed.), New York, Doubleday: 491–498.

Parodi, H., 1959, "Notice nécrologique sur Balthasar van der Pol, Correspondant pour les Sections des Académiciens libres et des Applications de la Science à l'Industrie", *Comptes Rendus de l'Académie des Sciences de Paris*, 249 (Octobre–Décembre): 1420–1422.

Poincaré, H., 1892–1899, *Les méthodes nouvelles de la mécanique céleste*, 3 vols., Paris, Gauthier-Villars.

Pol, B. L. (van der), 1919a, "The production and measurement of short continuous electromagnetic waves", *The London Edinburgh and Dublin Philosophical Magazine and Journal of Science*, 6th s., 38: 90–7.

Pol, B. L. (van der), 1919b, "A method of measuring without electrodes the conductivity at various points along a glow discharge and in flames", *The London Edinburgh and Dublin Philosophical Magazine and Journal of Science*, 6th s., 38 : 352–364.

Pol, B. L. (van der), 1919-20, "On the propagation of electromagnetic waves around the earth", *The London Edinburgh and Dublin Philosophical Magazine and Journal of Science*, 6th s., 38 : 365-382; and 40: 163.

Pol B. L. (van der), 1920, "A theory of the amplitude of free and forced vibrations", *Radio Review*, 1: 701–710; 754–762.

Pol, B. L. (van der), 1921, "Trillingshysteresis bij een triode-generator met twee graden van vrijheid", *Tijdschr. van het Nederlandsch Radiogenootschap*, 2: 125–142.

Pol, B. L. (van der), 1922, "On oscillation hysteresis in a triode generator with two degrees of freedom", *The London Edinburgh and Dublin Philosophical Magazine and Journal of Science*, 6th s., 43: 700–719.

Pol, B. L. (van der), 1924, "Gedwongen trillingen in een systeem met nietlineairen weerstand (Ontvangst met teruggekoppelde triode)", *Tijdschr. van het Nederlandsch Radiogenootschap*, 2: 57–71.

Pol, B. L. (van der), 1926, "On 'relaxation-oscillations'", *The London Edinburgh and Dublin Philosophical Magazine and Journal of Science*, 7th s., 2: 978–992.

Pol, B. L. (van der), 1927, "Forced oscillations in a circuit with nonlinear resistance", *The London Edinburgh and Dublin Philosophical Magazine and Journal of Science*, 7th s., 3: 65–80.

Pol, B. L. (van der), 1934 "The nonlinear theory of electric oscillations", *Proceedings of the Institute of Radio Engineers*, 22: 1051–1086.

Pol, B. L. (van der), 1946, "Music and elementary theory of numbers", *The Music Review*, 7: 1–25.

Pol, B. L. (van der), 1960, *Selected Scientific Papers* (H. Bremmer, C.J. Bouwkamp, eds.), 2 vols., Amsterdam, North-Holland.

Pol, B. L. (van der) and Mark, J. (van der), 1927, "Frequency demultiplication", *Nature*, 120: 363–364.

Pol, B. L. (van der) and Mark, J. (van der), 1928a, "The heartbeat considered as a relaxation oscillation, and an electrical model of the heart", *The London Edinburgh and Dublin Philosophical Magazine and Journal of Science*, (7) 6: 763–775.

Pol, B. L. (van der) and Mark, J. (van der), 1928b, "Le battement du cœur considéré comme oscillation de relaxation et un modèle éléctrique du cœur", *L'Onde Electrique*, 7ᵉ année, 81 (Septembre): 365–392.

Smith-Rose, R. L., 1959, "Dr. B. L. van der Pol (Obituary)", *Nature*, 184 (4692, October 3): 1020–1.

Volterra, V., 1931, *Leçons sur la théorie mathématique de la lutte pour la vie*, Paris, Gauthier-Villars.

Zeeman, E. C., 1970, "Differential equations for the heartbeat and nerve impulse", in: *Towards a Theoretical Biology* (Waddington, C. H., ed.), Edinburgh, Edinburgh University Press: 8–67.

Appendix: Two Unpublished Letters Written by Balthasar van der Pol to Vito Volterra

Dr. Balth. van der Pol

Jan Smitzlaan 12

Eindhoven

Holland

30-12-1929

Monsieur le Professeur V. Volterra

ROME.

Cher Monsieur,

C'est déjà depuis quelques années que je m'occupe aux recherches générales theoretiques et experimentales sur des phénomènes periodiques dans la nature, et je prends la liberté de vous envoyer, ci-enclus, quelques tirages à part de mes articles sur ce sujet.

Je sais, que vous avez publié plusieurs recherches importantes sur les solutions periodiques d'équations differentielles et intégrales. Parce que il est très difficile pour moi de trouver ces publications complètes, je vous serais très reconnaissant, si vous voudriez bien m'envoyer des tirages à part de vos travaux sur ce sujet. Pour le cas que vous n'avez plus de tirages à part, vous m'obligerez beaucoup, si vous voudriez bien me dire, où je peux trouver vos publications précieuses.

Veuillez agréer, cher Monsieur, avec mes vives remerciments, l'expression de mes sentiments les plus distingués.

Balth. van der Pol.

Natuurkundig Laboratorium

der

N. V. Philips'

Gloeilampenfabrieken

Kastanjelaan

TE Eindhoven

Eindhoven, le 28 janvier 1930

(Holland)

Monsieur le Sénateur

Professor Vito Volterra

Via in Lucina

ROMA.

Monsieur,

J'ai le plaisir de vous remercier beaucoup de l'aimable envoi de vos deux travaux, dont sûrtout l'article: "Variations and Fluctuations of the number of Individuals in animal species living together" a mon plus grand intérêt.

Permettez-moi de faire une petite observation, (faite avec l'intention de prédire des nouvelles possibilités futures de vos équations différentielles non-linéaires), en connection avec le cas, traité par vous sur pag. 49, où la "coefficience of increase" ε_r montre des petites fluctuations périodiques.

Nous avons constaté, pendant les dernières années théoriquement de même qu'expérimentellement, qu'une petite ride periodique avec une période , soit qu'elle parait dans un des coefficients d'une équation différentielle non-linéaire de deuxième ordre, soit qu'elle paraît comme terme libre dans le deuxième membre (comme force extérieure) peut avoir une influence fundamentelle sur le cours des phénomènes.

Aussi loin que, quand la période ω_1 avec laquelle le système oscillera, si les coefficients ε_r sont constants, est dans le voisinage de ½ ω_0, ⅓ ω_0, ¼ ω_0 ...., le système synchronisera automatiquement avec cette harmonique inférieure de ω_0 (dites $\omega_0 /$ n qui est le plus près de ω_1.

Dans le cas, traité par vous, où vous contemplez la vie ensemble de deux specimens, où l'un sert comme nourriture de l'autre, ceci reviendrait au suivant:

Quand l'oscillation libre du système avec des coefficients constants ε_r aurait une période de, par exemple 3¼ ans, alors, quand les ε_r's sont une petite ride avec une periode de precisément 1 an, le système total choisira automatiquement une période d'exactement 3 ans.

J'espère que cette observation vous intéresse, et si elle donnerait lieu à une observation sur ce sujet de votre part, il me serait très agréable de recevoir vos nouvelles.

Avec mes remerciments réitérés pour l'envoi de vos travaux si importants, veuillez agréer, Monsieur, l'expression de mes sentiments les plus sincères.

Dr. Balth. van der Pol.

P.S. J'ai encore le plaisir de vous faire parvenir par la présente un article: "Frequency Demultiplication" de M. van der Mark et moi, le sujet duquel est intimement lié à ce que nous avons traité dans cette lettre.

Annexe.

4 Transferring Formal and Mathematical Tools from War Management to Political, Technological, and Social Intervention (1940–1960)

Amy Dahan and Dominique Pestre

In characterizing the 1940s and 1950s, historians of science and technology have often stressed the importance of the university-military-industrial complex. In this framework, science (and especially the physical sciences) has generally been considered as the producer of a (disciplinary) knowledge that was relevant to technology and led to the development of material devices (like lasers) or military instruments (the Bomb or radar). Scientific milieus were at the core of the great East-West conflict known as the Cold War, and the profound changes in their material and cultural conditions have been emphasized. The effects of this new alliance on knowledge itself (for example in theoretical physics) have been precisely documented, and the centrality of instrumentation and production of all types of material devices underscored. By insisting on hardware, technological systems, and military "gadgets", these analyses have shown their great significance. No war, to be sure, is ever won by technology and science alone, but in the 1940s and 1950s they have played a role that can hardly be overstated[1].

In this essay, we have another goal. Our focus is not this set of practices and their consequences, but rather the conceptual tools that another group of "engineers-scientists" has during the same period perfected for the industrial management of war and Cold War, as well as the generalisation of these tools to other domains (from economics to logistics, from psychology to the integration of complex systems). These tools mobilized logic, mathematics, statistics and probability theory, signal theory, modeling, and the nascent computer. The most famous among them are known as operational research, game theory, general systems theory, linear and dynamic programming, queuing theory, cost-benefit analysis. They formed a set of techniques designed for optimizing military operations (such as nuclear strategy), the management of innovation, and the control of manpower. This set of practices derived from World War II experiences, from operational research, and from the work of several groups and networks mobilized through the Applied Mathematics Panel. During the Cold War, such tools and their applications were both on conceptual and practical levels developed and extended to the management of large technoscientific systems such as SAGE, the huge detection system set up over North America in the 1950s in order to forestall a possible attack by the Soviet Union (Hughes 1998, Edwards 1997). These practices also found their expression in some rational *calculation thinking* characterized by its hegemonic, universal vocation. "To conceptualize, design, and optimize all sorts of military, political, and social interventions" could be the motto for these new devices. Just like the physicists' complex, the center of this network was structured by the close interaction between high-level military hierarchies, large aeronautical and tele-communication corporations, and the scientific establishment[2].

Except for a few exceptional characters like John von Neumann (1903–1957) and Nobert Wiener (1894–1964), the traditional disciplinary intellectual histori-ography has tended to neglect the men who forged these tools. Several reasons can account for this. First, the unity of such practices and tools is hard to capture, as practitioners could be applied mathematicians, statisticians, economists, or spe-cialists of game theory or operational research. Second, while highly mathemati-cal, these domains were developed at the margins of university milieus and had little impact on pure mathematics. Finally, in the mid- to long-term, their effect was most strongly felt on economics, organization science, the social and cognitive sciences, and at the interface of these disciplines.

The present study has four parts. In the first, wartime groups occupied with operational research are briefly surveyed. The second part describes the dynami-cal development of new mathematical tools stemming from these networks and from the core techniques explored in the context of wartime emergency. In a third part, we focus on one of the especially crucial social sites for the fine-tuning of such techniques in the 1950s – the Research and Development Corporation (RAND Corporation). Finally, stepping back, we make a few comments on the characters of these new conceptual, and social, ways of doing things.

4.1 Operational Research and Mathematicians' Mobiliza-tion in World War II

In 1940, with the threat of a possible German landing, Britain urgently needed effective aerial defense. It was from this context of absolute urgency that opera-tional research emerged. Survival being at stake (literally, especially from May 1940 to May 1941), mobilization was general. The idea was to "hold". And for this, the use of current, but limited, means had to be optimized[3].

Operational research first confronted scientists with the task of turning the coastal radar chain protecting Great Britain into an operational system. As a first step, the electronic apparatus and the human system managing the information it produced had to be made to work. The absolute priority was to achieve maximal efficiency from the part of a system that included radar and ground observers, had to manage the detection and centralization of data, and determine the optimal use for fighter planes and air defense artillery. The obvious way to deal with the prob-lem therefore was to adopt a *systemic* approach.

In 1941, with the Germans unable any more to contemplate a British landing, operational research was shifted to other strategic locations, such as Coastal Com-mand Headquarters and the Navy. The best possible use of new weapons and tech-nologies that were regularly invented remained a concern. But the optimization of military operations (like the organization of anti-submarine warfare), the logistics of increasingly gigantic armies scattered all around the globe, and the apportionment and semi-planning of industrial production progressively became its most significant contributions. Keywords now were resource optimization (through calculation) and the increase in battle efficiency by deriving new rules for tactical behaviors from such studies. Since it provided political and military officials with recommendations that were judged efficient, operational research quickly spread to all General High Quarters.

The only example we will here mention concerns the analysis of escorting warships for North Atlantic convoys in March 1943. This study took into account all available statistical data from the previous two years. For each convoy, the analysis examined the number of escorting ships (from 1 to 15) and the size of the attacking group of submarines (from 1 to 20), and focused on correlations: the number of ships sunk as a function of escort size and submarine groups, the number of submarines sunk as a function of escort size, etc. The results were then summarized in the form of concrete recommendations (dictating optimal sizes for convoys and escorts), established on the basis of submarine tactics, attack procedures, the number of available ships, etc. This study was further completed by specific studies on convoys' spatial layouts (the number of lines and columns), the placement of escorts as a function of course followed, or the difference between day and night.

Let us note that this type of analysis relied on a wealth of extraordinarily detailed sources. From 1939 onward, each convoy indeed led to the production of several reports by the Navy. Produced by the thousands, these reports show how much this war (also) was from the start a paper war which relied on people to write and register the minutiae of battle narratives, people to inquire for information (aerodromes and maintenance centers were regularly visited to learn procedures in detail), and people to synthesize reports by means of statistics[4]. In parallel, experiments were conducted with the aim of optimizing the human factor in operational systems. These experiments concerned individuals (physiological and psychological experiments were for example conducted in airplanes in order to increase pilots' efficiency in spotting submarines visually) or collectives (by spatially remolding control rooms or by experimenting on operators' diction in view of increasing the reliability of information transfer and the speed of its treatment)[5].

In the American case, operations research equally infused headquarters, with however a few distinctive traits that would blossom after the war. Besides what was described above, American operations research was characterized by a high level of technicality and a more frequent recourse to modeling. Two types of reasons can be suggested to account for this difference. On the one hand, many American mathematicians were drafted into the war effort via the Applied Mathematics Panel (AMP) directed by Warren Weaver (1894–1978). On the other hand, the geopolitical situation of the country was very different: it was not directly submitted to the German menace and, much more than for Great Britain, it was that the planet as a whole that formed the theater of US forces. Planning a war waged thousands of miles away, scientists who served as advisors to them devised more global and more "theoretical" schemes.

Weaver and Vannevar Bush (1890–1974) had designed the AMP as a mathematicians' organization in charge of providing the military and other scientists mobilized in the war effort with mathematical assistance. It worked by contracts and involved eleven universities[6], as well as laboratories such as the Radiation Laboratory at MIT, Los Alamos, and Aberdeen Proving Grounds. The questions they tackled came under three headings and gave rise, in the immediate postwar period, to many research initiatives. The first set of questions stemmed from fluid mechanics, the study of shock waves, explosion theory in air or water, and submarine ballistic (torpedoes), as well as associated numerical problems. It also concerned the mathematical problems of analysis and probability theory raised by nuclear reactions (e.g., branching process). After the war, this domain dealt with problems at the confluence of hydrodynamics, computer, and numerical analysis. The second set of questions arose from operational research itself and logistic problems, from various statistical problems (tied to anti-aircraft defense or war material for example) and economic problems (resource management); it led to a manifold of disciplines and activities federating linear programming (and nonlinear, later mathematical, programming), game theory, decision-making theory, mathematical statistics, etc. The third set could be labeled as cybernetics. It stemmed from works on some defense systems involving men and machines, but also from an attempt at logically modeling the brain.

Concerning men, problems, and research themes, these three sets were never stable nor independent from one another. Over the 1940s and 1950s, they were in constant motion and reconfiguration. Some themes and results were shared, or could be transferred from one domain to the other. Personalities were to be found in various sets[7], and von Neumann was a major figure in the three of them. In the following, we will focus on the last two[8].

Within the AMP, some groups and institutions have played a more important role than others[9]. Among these, the Columbia Statistical Group seems to have been, according to W. Allen Wallis (b. 1912) who acted as research director, the most extraordinary group of statisticians ever gathered whether in number or in quality. It offered advice on the most varied statistical or probabilistic problems (fire control, bombing, statistical analysis of data). It gathered statisticians and economists, like Wallis, George B. Dantzig, Abraham Wald, Leonard Savage (who would partake in several Macy cybernetic conferences), the mathematician John Williams (later to head the mathematics department of the RAND Corporation), economists trained in statistics and probability theory, such as Harold Hotelling and Milton Friedman (later to receive the Nobel Prize for Economics), and engineers like Julian H. Bigelow, who was close to Wiener. Because of Bigelow, they were already applying means of computation and, in some cases, game theory to tactical questions. A further example would be Tjalling Koopmans, who was attempting to model a situation in which, given some offer in some ports and some demand in others, a transportation plan had to be designed so as to minimize costs. Besides the Statistical Mathematical Group, one must also mention the Princeton network (with Albert W. Tucker, J. Tuckey, Merrill Flood, Melvin Dresher, Lloyd S. Shapley, all of whom later to be involved with the RAND) as well as MIT's Operations Research Group headed by Philip Morse.

4.2 Mathematical Tools for Managing Social and/or Complex Systems

Two disciplinary domains arose from the work of groups contracted by AMP. Developed in the immediate after-war, they gave rise to many ramifications. The first of these domains, linear programming, formed one of the most significant mathematical domains stemming out of operational research; it was the natural extension of US Air Force planning activities (which required a vast coordination between material and personnel transportation capabilities and production and storage capabilities). Having undertaken rational studies of logistics for AMP, George Dantzig (b. 1914) was asked, after the war, to consider the mechanization of this process. He noticed that the goal of these complex procedures could be achieved by using inequalities rather than equations, and, at the end of 1947, he gave a mathematical description of the problem: find the solution of a system of linear equations or inequalities maximizing a given linear form (in the case of "linear programming," but this was only the first chapter of a vaster domain called "mathematical programming"). The "simplex method", used by Dantzig to solve the problem, reduced the study of inequalities to the search of a vertex on the convex polyhedron representing the set of possible solutions. Making the first computation by hand, Dantzig already had its mechanization on an electronic machine in mind. At once, was a large field of applications revealed: problems of transportation, distribution, personnel allocation, search for equilibrium between production and storage, etc. In 1949, being highly interested in applying linear programming to its logistical operations, the Office of Naval Research founded (and generously funded) the Logistic Branch of the Mathematical Sciences Division. It was headed by Fred Rigby, and produced a journal, the *Naval Research Logistic Quarterly*. Both of them were to play prominent role in later developments.

Related to the above, a second important theme of research emerged from the Statistical Group: sequential analysis and, more generally, statistical decision theory. The inspection of military equipment – a process by which items tested were rendered useless – had prompted research on algorithms and their tests on the basis of statistical samples. Simultaneously, the problem of statistical decision emerged: in the above example, one is always faced, in the decision process, with the alternative between testing further samples or stopping the process. Presented to Abraham Wald (1902–1950) by Friedman and Wallis, the problem received its first classified results as early as 1943. In 1945, Wald developed the "sequential probability ratiotest" (SPR), solving in particular the problem of determining whether a lot possessed a distribution A or a distribution B with given error risks while at the same time minimizing the number of items tested. This result would be immediately applied to quality control in industry[10]. Published in 1947, Wald's *Sequential Analysis* could be read from several points of view: probability theory, statistics, applications, etc. Its presentation was axiomatic: a particle making its way step by step in a "random walk," this process ending up when the particle went though a certain number of barriers and leaving a given interval. Very concrete concerns were expounded, like the elaboration of inspection plans insuring a continuous production with fixed quality standards (Wald and Wolfowitz 1945).

In parallel, being the concern of increasingly many mathematical works linked with statistics and probability theory, the theme of decision was itself widened: how can a final decision be derived from a random process selecting from various possibilities according to specific probabilities? Once again, let us underscore Wald's contribution (but here Savage should also be mentioned[11]), who followed a path that is very symptomatic of the preoccupations of this network of mathematicians: he indeed formulated several problems (including some concerning sequential analysis) in terms of *risk* (the risk could be that of not being within fixed margins of error) and the question then was to find a decision function that minimized risk. In *Statistical Decision Functions* (1950), Wald interpreted his theory of decision and statistical estimate in terms of von Neumann's game theory. Both conceptual frameworks were linked through a dictionary, the decision problem being reduced to a "game" between an experimenter (the statistician choosing rules for decision making) and "Nature" (which itself knew the random distribution that remained inaccessible to the player). Of course, Wald was forced to extend von Neumann's game theory to the case where an infinite number of strategies were possible and, using abstract probability spaces, to put it in a new mathematical framework. In fact the translation was never fully achieved. But Wald and Dantzig went on working together on the links between randomization in statistical problems and game theory.

Dantzig's research program was pursued in the nonlinear case by Albert Tucker (b. 1905). Passionate for combinatorial topology, Tucker was close to Solomon Lefschetz (1884–1972) and would succeed him as chair of Princeton's department of mathematics. In a seminar on game theory, he was then exploring the equivalence, conjectured by von Neumann, between two-person zero-sum games and linear programming. After having suggested a way of linking nonlinear logistic problems of transportation and distribution in electrical networks with the Kirchhoff-Maxwell law, Tucker was put in charge of a research project on linear programming, its foundations, and its links with game theory by the Office of Naval Research. Completely changing research interests, Tucker would for the next twenty years be concerned with this project – first in collaboration with his students Harald Kuhn (b. 1925) and David Gale[12]. Soon, a first important result establishing the conceptual, formal links with game theory was presented at a Chicago conference and published under the auspices of the Cowles Commission by Koopmans (Gale, Kuhn, and Tucker 1951). Published at the same time, Kuhn and Tucker's (1950) mathematical theorem specified the existence conditions for optimal solutions in a nonlinear programming problem. Various works were moreover devoted to solution algorithms for these problems that often involved a very large number of variables.

A new intellectual crossroad, this field was developed in the directions of optimization mathematics, the management sciences, and game theory. Here the interaction with economists became crucial. They had already encountered some aspects of linear programming, in particular Leonid V. Kantorovich (1912–1986), as early as 1938 (Kantorovich 1939) and Tjalling Koopmans (1910–1985), in 1942 in relation to a cargo transport problem raised by the Navy (Koopmans 1970). Koopmans – succeeding Jakob Marschak at the head of the Cowles Commission in 1948 – applied Dantzig's result to economics in the course of elaborating the theory of resource allotment. In an AMP contract, the mathematician Merrill Flood (b. 1908) had already formalized, in the case of air bombardment, one of the first examples of cost-benefit analysis. In the 1950s, cost-benefit analysis became common and were studied by Dantzig and others. From 1949 to 1962, symposia on mathematical programming were regularly held, fostering the development of a new field at the interface of mathematics, economics, and game theory[13]. In the 1960s, a society for mathematical programming was founded by Dantzig, Tucker, Koopmans, and a few others among whom was Kantorovich. Its paradigmatic figure was Dantzig, whom his colleagues described as a "mathematics-economic-computer engineer." Many also believed that he should have shared the Nobel prize jointly awarded to Kantorovich and Koopmans in 1971.

We have mentioned game theory. In fact, several contemporary studies underscored the deep change of status that game theory underwent during the Cold War, and at the RAND in particular. For von Neumann, a game (under its normal guise) was a model describing the interdependence between the players' *strategies* (i.e. the actions that they could freely choose to undertake) and their *utilities* (their preference or taste) which they had to maximize. Game theory therefore was an analysis of strategic activities among actors assumed *rational* – in the sense that they played solely to maximize their utilities – and placed in situations of uncertainty and randomness (bluff, lack of information, and so on). In his *Theory of Games and Economic Behavior*, written with Oskar Morgenstern (1902–1977) in 1944, von Neumann's project was to use this theory to formalize the behavior of economic agents; in a sense, the book questioned the ruling paradigm in economics according to which individual initiatives and limits within individual behavior were constrained were postulated. As a matter of fact, economists were quite reluctant to accept the new theory, Paul Samuelson (b. 1915) in particular criticizing it violently. On the contrary, the book was enthusiastically received by mathematicians. As was just noted, two theories *a priori* distinct from game theory – that is, statistical decision theory and linear programming – adopted its conceptual framework. At least among these mathematicians, game theory therefore appeared as the fundamental paradigm providing general concepts, method, and an anthropologic vocabulary that was flexible, efficient, and very suggestive.

In the course of the 1950s, a vast field of practices and disciplines was therefore constituted whose explicit vocation was to mathematize questions related to social phenomena, to rationalize organization and management procedures, in brief, to rationalize human behavior. This composite field encompassed mathematical theories proper or in the course of acquiring autonomy (such as the mathematical theory of optimization), mathematical branches specific to various domains of economics (cost-benefit analysis, resource allocation), polymorph mathematical theories and methods (mathematical programming), and finally a true theoretical framework, that is, game theory, whose status still remains to be more carefully examined here. In this disparate field, the major common notion was *optimization*. To optimize defense through the better use of available means had during the war been the concern of operations research and logistic; to maximize utilities or benefits, to minimize risks or costs, to optimize resource allocation were postwar goals. The concept of optimization permeated all results and theories mentioned above. In uncertain situations (random processes), optimization remained the great problem and was tightly connected to the new statistical theories of decision-making and control. Lastly, in game theory and all associated theories, *rational* behavior was always identified with optimizing behavior.

A last crossroad, the intellectual and scientific networks that were called the cybernetic group, whose origin also lay in World War II, emphasized analogies between man (nervous systems) and artificial machines. It will not be developed here but merely mentioned. The cybernetic group resulted from the convergence of two concerns. One focused on the idea that could be summarized by the meta-phorical shortcut according to which the brain functioned as a computer. Starting from the 1943 paper by Warren McCulloch (1898–1969) and Walter Pitt (b. 1923), it was furthered by von Neumann's (1948) semi-reproductive automata theory[14]. The other concern of the cybernetic group insisted on means of communication and control, on the notions of feedback and message, in systems involving men and machines – a distinction that, according to these notions, was no longer pertinent. It was first spectacularly expressed in the article "Behavior, Purpose, and Teleol-ogy" by Wiener, Arturo Rosenblueth (1900–1970), and Julian Bigelow (b. 1913) Starting from this core set of people among whom mathematicians Wiener and von Neumann played essential roles, the project of holding Macy Conferences on cybernetics was put together. From 1946 to 1953, they would gather anthropolo-gists (Gregory Bateson, Margaret Mead), biologists, psychiatrists and neurologists, and specialists from other domains in the human and social sciences.

4.3 An Emblematic Site for the Deployment of these Tools: The RAND Corporation.

We will not here recall the nature of the RAND Corporation and the conditions in which it was set up[15]. Let us just say that the RAND was the first postwar "think-tank". Put in place by the US Air Force, it most importantly recruited from groups associated with the AMP among whom there was many mathematicians and logicians. In intellectual terms a central role was played there by von Neu-mann. After the 1948 reorganization, the RAND was mostly known for its work on the creation of a "general science of war". Established in a single building at Santa Monica, California, whose architecture had been supervised by John Wil-liams himself, the RAND conceived of itself as an elite institution submitted to strict surveillance (since strategic questions were discussed and classified material accessible). Its collective model was that of high-level scientific experts developing new revolutionary techniques whose individual ideal was conceptual excellence and whose goal was the rationalization of decision-makers' measures. Concerned with the preparation of conventional warfare (logistics, transportation, disposition of forces), the conceptualization of nuclear warfare, individual preparation for commanding positions, and research (development, economic prevision, management), the RAND developed new conceptual tools often in close association with the use of the first computers.

Looking more precisely at the activities of the RAND from 1946 to the end of the 1950s, one notices a rather wide array of works. Some extended the methods of operations research to the conception of very large systems (such as the whole of US air defense). At the start of the 1950s, 37 groups dealing with this issue were considering questions as various as future radar development, the nature and geographical distribution of intercepting airplanes and missiles, vulnerability and nuclear warheads, the management of computing centers, the possibility of multiple hits, etc – betraying the wish, by varying all parameters, to provide a scientific assistance resting on rational decisions. This type of very general, all-encompassing "system analysis" however led only to disastrous results. True, it fostered the development of interesting, useful formal tools, but the complexity of the object of inquiry was such that concrete recommendations were often conflicting with common sense – and simply discarded. In 1951–52, a series of limited, less ambitious studies would in fact leave a stronger impression. Directed by logician and mathematician Albert Wohlstetter, who had worked on quality control during the war, these studies truly established the RAND's status as an expert organization for the military (Digby 1989, 1990).

As Wohlstetter explained, the object of the study was not to determine the best airbases for the US Air Force but rather to identify and situate the problem on the basis of strategic and economic considerations. More precisely, the choice of sites (a preliminary list of 13 sites as well as a set of 100 targets were selected) was studied with the aim of reducing costs in function of strategic considerations (e.g., comparing expected results of air vs. naval strikes), political considerations (agreement with allied countries to host US airbases and its dependability, quality of local workforce, possibility of preventive strikes), considerations taking into account the nature of sites (intrinsic costs associated with locating bases in the Antarctic), the nature of airbases (mere provisioning centers or airplane bases), various crucial distances (to targets, sites of arm-system supply, penetration distance through enemy defense, distance from enemy's strike sites), cost of initial installation and maintenance, possibility of air or ground supply, etc. The first phase of this study consisted in the precise identification of the most pertinent parameters and their links with one another (numbering in the thousands) so as to be able to study subsets more precisely. For example, how do other parameters (such as airplane or missile types needed for fulfilling missions) change when situations change (enemy's progressive withdrawal, for example, or damage on landing strips due to enemy bombardment)? These studies led to local recommendations (estimates for repairing time, for example, or suggestions for American airbase topography). Carried out on the basis of available data (airbases situated in Birmingham, Tokyo, or Cairo, identified targets in Moscow or Leningrad, known, expected, or desired material, costs provided by competent services), these studies were presented in the form of comparative tables and graphs. These results in turn brought about further simplifications (it was for example concluded that some functions that bases could fulfil had to be separated and individually evaluated) and wider conclusions. Thus it was concluded that it would be better, for a variety of reasons, to maintain only supplying bases for airplanes, which should all be repatriated and kept on alert in the United States – a suggestion that was adopted by the Air Force (Wohlstetter 1951, 1952).

Also popular at the RAND Corporation, applications of game theory concerned the tracking down of targets, the timing of missile launches, air warfare tactics – and, of course, the conceptualization of nuclear warfare. In 1950, the RAND numbered more than 200 mathematicians under the supervision of Williams, a former member of the Columbia Statistical Group. The mathematical division was called "Evaluation of Military Worth". Nearly all game theorists were on its pay-roll, either as permanent staff or contractors. Two new directions would become predominant. Called "differential game theory," the first was globally concerned with military applications – to compute the optimal strategy of a missile chasing a bomber, or, more generally, to direct an object whose kinetic parameters were controlled towards another moving object with its own characteristics. Mathematicians like Richard Bellman (1920–1984)[16] were heavily involved in this active field that mobilized various applied mathematics results from differential and numerical analysis. One must here underscore the mobilization of game theory's vocabulary and conceptual framework for the formal analysis of the very same (feedback) system that had been the paradigm for Wiener's cybernetics and for some nonlinear mathematical physics problems (solved as such by Lev S. Pontryagin (1908–1988) in the Soviet Union with no reference to either game theory or cybernetics, see chapters 8 and 9). Instigated and promoted by the RAND's military and political leadership, the second use of game theory was to think formally about the strategic conflict against the Soviet Union and intercontinental nuclear warfare. It was, for that matter, during a seminar at the RAND that the famous model of the prisoner's dilemma was invented (and christened by Tucker) as an illustration of the conflict between individual benefit and collective choice. It was in the same setting that the model was later used to explore the notion of *rational choice*, an axiom of game theory.

Three remarks must be made with respect to game theory and the vogue it enjoyed at the RAND. First, this vogue was paradoxical since, notwithstanding the skill of its practitioners, game theory would soon be discredited at the RAND for delivering less than expected and, in particular, for lacking as an efficient science of war. Second, the use of game theory was highly pragmatic since it was constantly tested on games actually played around a table by members of the RAND. For Flood, for instance, axiomatic structures developed in a game-theoretic framework had to be submitted to controlled experimental situations so as to evaluate their practical pertinence. Finally, strategic games between teams replaying scenes from the last war or forecasting the next were omnipresent at the RAND. As in war schools, these games were played quickly or on several days, using historical examples or simulations constructed on the basis of models and theories and "running on computers". In the following, we shall come back to the special importance of these games (in the dual sense of theoretical models and actually played strategic games) for sociability at the RAND, its subculture – and the benefits taken from them:

enjoyment derived by these men (there was no women in this world), certainty of being "scientificizing" war and revolutionizing its conceptualization, etc.

Lastly, let us mention other activities valued by the RAND: (1) the development of linear and dynamic programming techniques (the RAND would thus massively contribute to the development of the "software" needed for the SAGE system); (2) the production of an economic theory and techniques relying on the same mathematical tools; and (3) the development of an experimental psychology aiming at optimizing and improving collective behaviors (first in SAGE's control centers, in a Latin-American village by the end of the decade), more globally leading to the construction of a new approach to the social sciences.

4.4 On a Few Characters of these New Scientific Modes

Before we characterize these new scientific modes, let us remark that several "cultures" shaped the mathematicians/engineers/scientists who deployed them. The first of these – the most immediate and most decisive – was acquired during the war and mobilization years. In his study of Wiener and Bigelow's work on the Anti-Aircraft Predictor, P. Galison underscored the influence of war and weapon design on the cybernetic conception of the system "man-airplane-radar-predictor-artillery" as a single system in which the human component can be replaced by a machine and conversely. The enemy is a cybernetic entity; war is a conflict between cybernetic entities. Similarly, it seems clear that the development of game theory and the conceptualization of the cold war went hand in hand. Coupled with this was an engineering culture whose values were very strongly felt in people like Wiener (as early as the 1930s), the designers of electric network or information theory (Claude Shannon (1916-2001)), or the specialists of dynamic programming (Bellman). In his introduction to *Cybernetics* (1948), Wiener underlined how much he was determined by the close contacts he had with Vannevar Bush's computing machine development program and his collaboration with Yuk Wing Lee around the conception of electric networks. Around hardware and computation problems, in particular, this engineering culture was greatly extended in the war and immediate postwar contexts. A cultural tradition related to logic proper was also important – for von Neumann and the Princeton mathematicians, for theoretical statisticians and economists, for the cybernetic group (and in particular McCulloch): logic and logical proposition calculus stemming from Russell's and Whitehead's *Principia mathematica* (Bertrand Russell (1872-1970), Alfred Whitehead (1861-1947)), Hilbert's and Gödel's work (David Hilbert (1852-1943), Kurt Gödel (1906-1978))on the arithmetization of logic and the logicization of arithmetic, Turing's and Church's studies of computability (Alan Turing (1912-1954), Alonzo Church (1903-1995)). The logic culture was moreover influential via logical positivism. In short, the impact that the immense body of work done by mathematicians and logicians in the 1920s and 1930s had on the scientists and engineers of this group cannot be underestimated[17]. Finally, Hilbert's axiomatic and formalist ideal was the foundation upon which everyone agreed even though contemporary pure mathematicians' excesses along this line fostered criticism and rejection.

As far as work was concerned, a first common trait should be underlined – that is, the claim for transdisciplinarity or a-disciplinarity. Mathematical tools and modeling, it was believed, made it possible to transcend disciplinary boundaries and tackle human and technological systems globally. Economists, engineers, mathematicians, psychologists, or physicists came from various backgrounds, but coherence was cemented by their mastery of formal logical tools which came with a value scale of their own. Working in different spaces (universities, "think-tanks", military laboratories, industrial companies such as Bell or Douglas Aircraft) or circulating among them, they had together become familiar with these new modes during World War II. They were often quite young. At the RAND, for example, the average age was around thirty in 1950 – surely a sign of the enterprise's intellectual novelty, and perhaps also of its radical character (Bowker 1993, Digby 1989, 1990). This kind of extra-disciplinary knowledge was aimed at understanding the Whole. Intricate questions were tackled globally and decision optimized, or so its practitioners liked to believe. Armed with their new knowledge and their new approaches, they felt capable of treading new scientific ground by introducing unfamiliar tools (like modern algebra) or approaches (like the construction of models for maximization). The rationality that was offered was algorithmic and mechanical (as in a computer) and was presented to political, military, and industrial decision-makers as having a great utility value[18].

These approaches and work principles had been recomposed (or acquired their legitimacy) by the new social links woven by the cold war. True, practices and theories had their own intrinsic logic (there were progressions leading such scientist to take up such modeling techniques, or to generalize such mathematical technique with the aim of addressing such specific problem), but, like for physicists, the new social and political situation (to work for the Admiralty or the RAND, to work at solving problems posed by the outside, to be involved in specific social networks, in short, to be a component of the new military-industrial-academic alliance inherited from the years 1939–1945) brought about, induced – and then legitimized – these approaches, the type of solutions chosen, and, in the end, the type of science produced. Granted that in his individual practice a scientist could feel that he was only developing a well-circumscribed technique (to link formally an aspect of game theory with linear programming techniques, for example), taken as a whole this collection of works has largely been allowed for by the "context"; and, for scientific milieus as well as society in general, they themselves gave rise to a new, legitimate way of "doing well".

What about the norms and values which authorized (or based, or eased) the conduct of such work, and those which, in turn, were diffused into the social body? Modeling frameworks such as game theory which were built for the understanding of idealized, but nevertheless very concrete, situations, we want to argue, gave rise to striking images outside of groups formed by specialists. Formally, axioms on which models were based only had local validity (von Neumann and Morgenstern's definition of "rational behavior", for instance). However, the vocabulary used being commonplace (and sometimes taken from common psychology), and the statements produced being so forth applied to ordinary life (or at least, perceived as such by many, including their promoters) and socially sanctified as being made operative by scientists – all this had a wide social impact. This produced new ontologies of human beings in society, and new discourses on their "normal" behaviors. As an analogy, one can look at the construction of socio-professional categories in France. From a certain point of view, this elaboration truly was a scientific endeavor. But, from another, once defined, these categories were widely taken up and recycled by various social actors – for example in collective labor agreements. It consequently partook in the reorganization of society, it remodeled it, it directly *performed* the social order in accordance with its own values[19].

The worldview at the basis of this enterprise (and of the RAND in particular) has been expressed by Weaver as a contribution to the development of "rational life" (the recurring use of this qualifier with its strong moral implication is quite striking). Transportation systems, armies, cities, or societies – all technological and human systems were, according to this worldview, collectives of the same nature: coordinated, or susceptible of being coordinated, and optimized in function of common, shared principles. Reason (via mathematics) was to be the privileged tool for the emergence of "the best for everyone". We are less concerned here by the fact that mathematics can be efficient when problems are posed in such a way (to say this is rather banal, since this is exactly what mathematics knows well how to do) than by the "ideological" operation that the worldview captures. Compared to European "social sciences" in the 1920s and 1930s concerned with variety of interests, description of domination relationships and exhibition of various value regimes in society, postwar attitudes were geared toward the better coordination (and control) of society in view of a single goal portrayed as a matter of course and shared by all its members (at the time, it often was "national security"). While the posture might have had some relevance for the deployment of US Air Force airbases around the globe, or even for industrial management, when describing social or economic life, it acquired a normative power that must be seriously questioned.

The approaches and theoretical frameworks discussed here therefore carried worldviews resulting from a world shaped by the Cold War. This interdependence is revealed by specific social rites going on at the RAND. Besides strategic games prized and practiced by its scientists and the fact that they rarely took the doves' side, a few other social behaviors could be more closely examined. Let us think at those challenges that they threw at each other seemingly in a recurrent fashion, at those "high-alcohol, high-IQ" parties organized by Williams and von Neumann and mentioned by R. Leonard, at those big cars driven at a high speed, at that violence expressed in the human relations between scientific promoters of the SAGE system – betraying values, in brief, that were highly virile and imperialistic. This was a men's universe – men comfortable with playing at annihilating half of the world, men convinced that the best, more rational manner of guiding the world was expert counsels and enlightened princes, men confident in their own value and ignoring doubt[20].

4.5 Three Short Remarks as Way of Conclusion

This moment in history renewed with a mystique of science and technology believed to be able to solve all problems (labor problems as well as logistic ones) and win all wars (whether social or military). This was a mystique about "algorithmization" and numerical computation, coupled with generalized automation and elimination of human labor (a thesis perfectly illustrated by the history of machine-tools in the US). That these programs were not always successful (and this is a euphemism) was of minor importance. Only the direction of the arrow of progress mattered: THE project (Hughes 1998, Galison 1994).

This was also a moment characterized by the renewal of a managerial/scientific thought. Always directly articulated around operative purposes, this type of work was carried out in close interaction with the leadership, it dreamt of rational choice for action, of complete control and anticipation. It could therefore become a profitable substitute for the market's invisible hand and provide scientific means for managing scientific and technological inventiveness as well as production and quality control (Locke 1989).

This moment, finally, was a moment when new social sciences were invented to replace the human sciences simply dubbed inefficient: psychological theories and practices based upon experimentation and aiming at optimizing strategic actors' behavior under conditions of stress and emergency; models of the social world built on the basis of axiom systems and doing away with the critical approaches typical of pre-war European traditions, etc. It was often the modes of thinking of the engineers that were picked, with their stress on the optimization of social relation regimes on the basis of parameters taken as universally valid, and constituting both the goal and the justification for such studies.

Notes

1 See, in particular, Forman 1987, 1989, Galison 1988, Galison and Hevly 1992, Glantz 1978, Godement 1978, 1979, Hoch 1988, Kevles 1988, Leslie 1987, 1993, Mendelsohn 1988, 1989, Pickering 1995, Roland 1985, Sapolsky 1990 and Schweber 1986.

2 A list of reference about this question should include Heims 1980, 1993, Leonard 1994, 1995, (forthcoming), Edwards 1997, and Mirowski 1999.

3 Archival research for this section has been done at the British military archives (RAF, Coastal Command, and Admiralty archives). It will be developed in further publications. See also Leonard (forthcoming), Mirowski 1999, and Jolink 1999. On operations research in the United States, see Rau 2000.

4 Some mythical history has claimed that operational research began with the application of mathematics and statistics to the analysis of operations. The statistical techniques used were in fact rather rudimentary and much less important than presence on the ground and strict control over the constitution of data. After the war, stories about operational research were an important ideological and professional issue pitting ordinary engineers against biologists, physicists and mathematicians about who had started it, doing what, and coming from where.

5 For parallel (although different) approaches in the US concerning the integration of machine and man, command centers, communication with noise, etc., see Edwards's 1998 study on the Psycho-Acoustic Laboratory.

6 The list of 'Technical Representatives' contracted by the Applied Mathematical Panel is the following. *Applied Mathematics Group*: New York University, Richard Courant; Columbia University, Moulton, Saunders MacLane and Arthur Sard; Brown University, Richardson; Institute for Advance Study, von Neumann; Princeton, Merrill Flood; Northwestern, Moulton, Leighton; Carnegie Institute of Technology, Pasadena, Adams; Harvard University, Garrett Birkhoff; University of New Mexico, Workman. *Statistical Research Groups*: Columbia, Hotelling; Berkeley, Jerzy Neyman; Columbia, Schilt; Princeton, S. S. Wilks. *Computation*: Franklin Institute, Allen; National Bureau of Standards, Arnold Lowan.

7 For example, Wiener is involved in the second and third sets, Bellman in the first and second, and Savage in the second and third.

8 About the first set, see Dahan 1996, 1997, 1999.

9 Columbia's three groups alone amounted for half of total AMP funding.

10 At the end of the 1940s, the Bell Company was very interested by this question.

11 Savage wanted to reformulate statistical behaviorism or the statistical theory of decision-making (in situation of uncertainty) so as to infer satisfactory rules for acting. He started with two notions: utility function (providing a strict hierarchy for individual interests) and personal (or subjective) probability as opposed to Wald, Ronald A. Fischer (1890-1962) or statisticians from the frequentist school.

12 After having spent the war at the Radiation Laboratory in MIT, Gale came to Princeton for his Ph.D. and started working on game theory.

13 The symposia took place in Chicago (1949), Washington (1951, 1955), Santa Monica (1959), Chicago (1962). Each time, 30 to 50 papers were presented; to the slow growth in the 1960s, a true explosion ensued in the 1970s.

14 On the various steps in the construction of this analogy, see Dupuy 1993. Let us recall that when ENIAC was started at the More School of Engineering of the University of Pennsylvania, in 1943, one had yet to come up with the idea that the logical structure of a computing machine was separable from its electronic architecture constrained by technology and the nature of physical components. It was von Neumann, called on as a consultant by the builders of ENIAC, who formulated the idea. Facing a material object, he adopted the same attitude as McCulloch and Pitt facing the brain: a logical machine – in fact a universal Turing machine – was abstracted.

15 First conceived by Douglas Aircraft in 1945 as a lobby for military research, this organization became in 1948 a rather singular institution. Contracted by the Air Force, which was its only client, it was importantly sponsored by the Ford Foundation. On the RAND, see Hounshell 1997, Leonard (forthcoming), Mirowski 1999, Edwards 1997, Hughes 1998, Digby 1989, 1990, Hitch and McKean 1960, Kahn 1960, and Kaplan 1983.

16 After a few stays at the Rand from 1948 onward, Richard Bellman left Princeton for the Rand. He was, he said "a modern intellectual, using the results of his research to solve the problems of contemporary society". He would always remain enthusiastic about this place that allowed him to work simultaneously on systems analysis, numerical analysis, game theory, mathematical modeling for the most complex physical processes – in particular, the study of "branching processes" started by Stanislaw Ulam (1909–1984) – as well as the social sciences, etc. Developing his own field of research, Bellman called it "dynamic programming".

17 In London, Wiener was Russell's student; spending some time at Göttingen when the most decisive works on metamathematics and logic came out, von Neumann had himself worked on axiomatic set theory; via the philosopher Rudof Carnap (1891–1970) – a leading figure of logical positivism then at Chicago – McCulloch was very influenced by the logicist approach of Russel and Hilbert whom he had already studied; Turing, finally, wrote an article in 1936–1937 which became a must studied by all mathematicians. At Princeton in 1938, he defended a thesis on logic directed by Church (Lévy 1993).

18 On algorithmization and techniques that made it effective (like Dantzig's simplex method), see Locke's example (1989) concerning industrial management (pp. 18ff).

19 Von Neumann and Morgenstern 1972.

20 Leonard (forthcoming), Hughes 1998. Let us note that social rites were very different in the cybernetic group Heims 1993.

References

Primary sources

Bellman, R., 1984, *The Eye of the Hurricane. An Autobiography*, Singapore, World Scientific.

Bigelow, J., 1980, "Computer development at the Institute for Advanced Study", in: Metropolis, Howlett, and Rota 1980: 291–310.

Birkhoff, G., 1977, "Applied mathematics and its future", in: *Science and Technology in America: An Assessment* (Thomson, R. M., ed.), Washington, U.S. Dept. of Commerce, National Bureau of Standards.

Birkhoff, G., 1980, "Computing developments 1935–1955, as seen from Cambridge, U.S.A.", in: Metropolis, Howlett, and Rota 1980: 31–30.

Bode, H. W., 1945, *Network Analysis and Feedback Amplifier Design*, Princeton, Van Nostrand.

Bush, V., 1929, *Operational Circuit Analysis*, New York, Wiley.

Bush, V., 1931, "The differential analyzer. A new machine for solving differential equations", *Journal of the Franklin Institute*, 212: 447–488.

Dantzig, G. B., 1963, *Linear Programming and Extensions*, Princeton, Princeton University Press.

Dantzig, G. B., 1984, "Reminiscences about the origins of linear programming", *Memoirs of the American Mathematical Society*, 48(4): 1–11.

Duren, P., Askey, R. A., and Merzbach, U. C. (eds.), 1988–1989, *A Century of Mathematics in America*, 3 vols., Providence (R.I.), American Mathematical Society.

Gale, D., Kuhn, H. W., and Tucker A. W., 1951, "Linear programming and the theory of games", in: *Activity Analysis of Production and Allocation* (Koopmans, T. J., ed.) Cowles Commission Monograph, 13, New York, Wiley.

Kahn, H., 1960, *On Thermonuclear War*, Princeton, Princeton University Press.

Kantorovitch, L. V., 1939, *Mathematicheskie metody organizachii i planirovaniya proizvodstva* (Mathematical Methods of Organizing and Planning Production), Leningrad, S.U. Press.

Koopmans, T. J., 1970, "Exchange ratios between cargoes on various routes (Memorandum for the combined Shipping Adjustment Board, Washington D.C., 1942) in: *Scientific Papers of T. J. Koopmans*, New York, Springer: 77–86.

Kuhn, H. W., 1976, "Non linear programming: a historical view", *SIAM-AMS Proceedings*, vol. IX, Providence (R.I.), American Mathematical Society.

Kuhn, H. W. and Tucker, A. W., 1951, "Non linear programming", in: *Proceedings of the Second Berkeley Symposium on Mathematical Statistics and Probability* (J. Neyman, ed.), Berkeley, University of California Press: 481–492.

Kuhn, H. W. and Tucker, A. W. (eds.), 1950, *Contributions to the Theory of Games I*, Annals of Mathematical Studies, 24, Princeton.

Kuhn, H. W. and Tucker, A. W. (eds.), 1953, *Contributions to the Theory of Games II*, Annals of Mathematical Studies, 28, Princeton.

Lax, P., 1977, "The bomb, sputnik, computers, and European mathematics", in: Tarwater 1977: 129–135.

Lax, P., 1988, "The flowering of applied mathematics in America", in: Duren, Askey, and Merzbach 1988-89: vol. II, 455-466.

Marschak, J., 1950, "Rational behavior, uncertain prospects, and measurable utility", *Econometrica*, 18: 112-141.

MacLane, S., 1989, "The Applied Mathematics Group at Columbia in World War II", in: Duren, Askey, and Merzbach 1988-89, vol. III: 495-515.

McCulloch, W. S., and Pitts, W., 1943, "A logical calculus of the ideas immanent in nervous activity", *Bulletin of Mathematical Biophysics*, 5: 115-133.

Metropolis, N., Howlett, J. and Rota, G. C., (eds.), *A History of Computing in the Twentieth Century*, Academic Press, New York, 291-310.

Morse, P. M., 1948, "Mathematical problems in operations research", *Bulletin of the American Mathematical Society*, 54: 602-621.

Morse, P. M., 1977, *In at the Beginnings: A Physicist's Life*, Cambridge (Mass.), MIT Press.

Morse, P. M. and Kimball, G. E., 1951, *Methods of Operation Research*, Cambridge (Mass.), MIT Press and New York, Wiley.

Neumann, J. von and Morgenstern, O., 1944, *Theory of Games and Economic Behavior*, New York, Wiley (2nd ed., 1947).

Neumann, J. von, 1951, "The General and Logical Theory of Automata", in: *Cerebral Mechanisms in Behavior: The Hixon Symposium* (Jeffries, L.A., ed.), New York, Wiley.

Rees, M., 1977, "Mathematics and the government: the post-war years as augury of the future", in: Tarwater 1977: 101-116.

Rees, M., 1985, "Mina Rees, interviewed by Rosamund Dana and Peter J. Hilton", in: Albers and Alexanderson 1985: 256-267.

Rees, M., 1988, The mathematical sciences and World War II, in: Duren, Askey, and Merzbach 1988-89: vol. I, 275-289.

Rosenblueth, A., Wiener, N., and Bigelow, J. H., 1943, "Behaviour, purpose and teleology", *Philosophy of Science*, 10: 18-24.

Samuelson, P. A., 1938, "A note on the pure theory of consumer's behavior", *Economica*, n.s., 5: 61-71.

Savage, L. J., 1954, *The Foundations of Statistics*, New York, Dover.

Shapley, L. S., 1953, "A value for n-person games", in: Kuhn and Tucker 1953: 307-317.

Ulam, S. M., 1958, "John von Neumann, 1903-1957", *Bulletin of the American Mathematical Society*, 64 (3), part. 2: 1-49.

Ulam, S. M., 1960, *Computing Machines as a Heuristic Aid-synergesis. A Collection of Mathematical Problems*, New York, Wiley.

Ulam, S. M., 1976, *The Adventures of a Mathematician*, New York, Scribners.

Wald, A., 1945, "Statistical decision functions which minimize the maximum risk", *Annals of the Mathematics*, 46: 265-80.

Wald, A., 1947, *Sequential Analysis*, New York, Wiley.

Wald, A., 1950, *Statistical Decision Functions*, New York, Wiley.

Wald, A. and Wolfowitz, J., 1945, "Sampling inspection plans for continuous production which insure a prescribed limit on the outgoing quality", *Annals of Mathematical Statistics*, 16 (1).

Wallis, W. A., 1980, "The statistical research group", *Journal of American Statistical Association*, 75: 320–335.

Wiener, N., 1948, *Cybernetics, or Control and Communication in the Animal and the Machine*, New York, Wiley (2nd ed. 1961).

Wiener, N., 1956a, *I am a Mathematician: The Later Life of Prodigy*, Cambridge (Mass.), MIT Press.

Wiener, N., 1956b, "Nonlinear prediction and dynamics", in: *Proceedings of the Third Berkeley Symposium on Mathematical Statistics and Probability* (Neyman, J., ed.), Berkeley, University of California Press: 247–252.

Wohlstetter, A., 1951, *Economic and Strategic Considerations in Air Base Location: A Preliminary Review*, RAND report D 1114, December 1951.

Wohlstetter, A., 1952, *Campaign Time Pattern, Sortie Rate, and Base Location*, RAND report D 1147, February 1952.

Secondary sources

Albers, D. J. and Alexanderson, G. L. (eds.), 1985, *Mathematical People. Profiles and Interviews*, Basel, Birkhäuser.

Aspray, W., 1988, "The emergence of Princeton as a world center for mathematical research", 1896–1939, in: *History and Philosophy of Modern Mathematics* (Aspray, W. and Kitcher, Ph., eds.), Minneapolis, University of Minnesota: 346–366.

Aspray, W., 1990, *John von Neumann and the Origins of Modern Computing*, Cambridge (Mass.), MIT Press.

Balogh, B., 1991, "Reorganizing the organisational synthesis: federal-professional relations in modern America", *Studies in American Political Development*, 5: 119–172.

Bennett, S., 1993, *A History of Control Engineering 1930–1955*, London, Peter Peregrinus.

Bowker, G., 1993, "How to be universal: some cybernetic strategies, 1943–1790", *Social Studies of Science*, 23: 107–127.

Dahan Dalmedico, A., 1996, "L'essor des mathématiques appliquées aux Etats-Unis: l'impact de la seconde guerre mondiale", *Revue d'Histoire des Mathématiques*, 2: 149–213.

Dahan Dalmedico, A., 1997, "Mathematics in the 20th century", in: Krige and Pestre 1997: 651–667.

Dahan Dalmedico, A., 1999, "Pur versus appliqué? Un point de vue d'historien sur une 'guerre d'images'", *La Gazette des mathématiciens*, 80: 31–46.

Dahan Dalmedico, A., 2001a, "An image conflict in mathematics after world war II", in: *Changing Images in Mathematics. From the French Revolution to the New Millenium* (Bottazzini, U. and Dahan, A., eds.), London-New York, Routledge: 223–253.

Dahan Dalmedico, A., 2001b, "History and epistemology of models: meteorology as a case-study (1946–63)", *Archive for History of Exact Sciences.*, 55: 395–522.

Dahan Dalmedico, A. and Pestre, D., 1998, "Comment parler des sciences aujourd'hui", in: *Impostures Scientifiques* (Jurdant, B., ed.), Paris, Éditions La Découverte: 77–105.

Dahan Dalmedico, A. and Pestre, D., forthcoming, "Les sciences pour la guerre (1940–1960), Paris, Presses de l'EHESS.

Digby, J., 1989, *Operations Research and Systems Analysis et RanD*, 1948–1967, RanD N-2936-RC.

Digby, J., 1990, *Strategic Thought and RanD*, 1948–1963, RanD N-3096-RC.

Dupuy, J. P., 1993, "L'essor de la première cybernétique 1943–1953", in: *Histoires de Cybernétique*, Cahiers du CREA, 7, Paris, Ecole Polytechnique-CREA: 9–139.

Dupuy, J. P., 1997, *Genèse des Sciences Cognitives*, Paris, Éditions La Découverte.

Edwards, P. N., 1997, *The Closed World. Computers and the Politics of Discourse in Cold War America*, Cambridge (Mass.), MIT Press.

Forman, P., 1987, "Behind quantum electronics: National security as basis for physical research in the United States, 1940–1960", *Historical Studies in the Physical and Biological Sciences*, 18 (1): 149–229

Forman, P., 1989, "Social niche and self-image of the american physicist", in: *The Restructuring of Physical Sciences in Europe and the United States, 1945–1960* (De Maria, M., Grilli, and Sebastiani,F., eds.), Singapore, World Scientific: 96–104.

Galison, P., 1988, "Physics between war and peace", in: Mendelsohn, Smith, and Weingart 1988: 47–86.

Galison, P., 1994, "The ontology of the enemy: Norbert Wiener and the cybernetic vision", *Critical Inquiry*, 21: 228–266.

Galison, P., 1996, "Computer simulations and the trading zone", in: *Disunity of Science: Boundaring, Contexts, and Power* (Galison, P. and Stump, D. J., eds.), Stanford (Cal.), Stanford University Press: 118–157.

Galison, P., 1997, *Image and Logic: A Material Culture of Microphysics*, Chicago, The University of Chicago Press.

Galison, P., Hevly, B. (eds.), 1992, *Big Science, the Growth of Large Scale Research*, Stanford, Stanford University Press.

Glantz, S. A., 1978, "How the department of defence shaped academic research and graduate education", in: *Physics Careers, Employment and Education* (Perl, M.L., ed.), New York, AIP: 109–122

Godement, R. 1978–1979, "Aux sources du modèle scientifique américain", *La Pensée*, 201: 33–69; 203: 95–122; and 204: 86–110.

Goldstine, H. H., 1972, *The Computer from Pascal to von Neumann*, Princeton (N.J.), Princeton University Press

Gerowitch, S., 1999, "Striving for 'Optimal Control': Soviet Cybernetics as a 'Science of Government'", unpublished.

Heims, S. J., 1980, *John von Neumann and Norbert Wiener. From Mathematics to the Technologies of Life and Death*, Cambridge (Mass.), MIT Press.

Heims, S.J., 1993, *The Cybernetics Group*, Cambridge (Mass.), MIT Press.

Hitch, C. J. and McKean, R. N., 1960, *The Economics of Defense in the Nuclear Age*, Cambridge (Mass.), Harvard University Press.

Hoch, P. H., 1988, "The crystallisation of a strategic alliance: the american physics elite and the military in the 1940's", in: Mendelsohn, Smith, and Weingart 1988: 87–177.

Hounshell, D., 1997, "The cold war at RanD, and the generation of knowledge", *Historical Studies in the Physical and Biological Sciences*, 27: 237–267.

Hughes, T. P., 1998, *Rescuing Prometheus*, New York, Pantheon Books.

Israel, G., 1996, *La mathématisation du réel*, Paris, Le Seuil.

Israel, G., and Millán Gasca, A., 1995, *Il mondo come gioco matematico. John von Neumann scienziato del Novecento*, Roma, La Nuova Italia Scientifica.

Jolink, A., 1999, "The Travelling Salesman Returns from the War: Tjalling Koopmans and War-time Studies for Peace-Time Applications", unpublished.

Kaplan, F., 1983, *The Wizards of Armageddon*, New York, Simon & Schuster.

Kahn, H., 1960, *On Thermonuclear War*, Princeton (NJ), Princeton University Press.

Kevles, D. J., 1988, *Les physiciens*, Paris, Anthropos.

Krige, J. and Pestre, D. (eds.), 1997, *Science in the 20th Century*, London, Harwood Academic Publishers.

Leonard, R. J., 1995, "From parlor games to social science: von Neumann, Morgenstern, and the creation of game theory 1928–1944", *Journal of Economic Literature*, 33: 730–761.

Leonard, R. J., 1998, "Ethics and the excluded middle: Karl Menger and social science in interwar Vienna", *Isis*, 89: 1–26.

Leonard, R. J., forthcoming, "Into the Labyrinth: Social Science and the Present Danger, at RAND 1946–1960".

Leslie, S. W., 1987, "Playing the education game to win: the military and interdisciplinary research at Stanford", *Historical Studies in the Physical and Biological Sciences*, 18 (1): 55–88

Leslie, S. W., 1993, *The Cold War and American Science, The Military-Industrial-Academic Complex at MIT and Stanford*, New York, Columbia University Press.

Lévy, P., 1993, "L'œuvre de Warren McCulloch", in: *Histoires de Cybernétique*, Cahiers du CREA, 7, Paris, Ecole Polytechnique-CREA: 211–255.

Locke, R. R., 1989, *Management and Higher Education since 1940. The Influence of America and Japan on West Germany, Great Britain and France*, Cambridge, Cambridge University Press.

McFarland, S. L., 1995, *America's Pursuit of Precision Bombing, 1910–1945*, Washington, Smithsonian Institution Press.

Mendelsohn, E., 1989, "Science, technologie et modèles militaires d'interaction", in: Salomon 1989: 49–74.

Mendelsohn, E., Smith, M. R., and Weingart, P. (eds.), 1988, *Science, Technology and the Military*, Dordrecht, Kluwer.

Mirowski, P., 1991,"When games grow deadly serious, the military influence on the evolution of game theory", in: *Economics and National Security, A History of their Interaction* (Goodwin, C. D. W., ed.), Duke University Press: 227–255.

Mirowski, P., 1999, "Economics meets operations research in mid-century", *Social Studies of Science*, 29: 685–718.

Mirowski, P., forthcoming, *Machine Dreams*.

Nasar, S., 1998, *A Beautiful Mind. A Biography of John Forbes Nash*, New York, Simon & Schuster.

Owens, L., 1989, "Mathematicians at war: Warren Weaver and the Applied Mathematics Panel", in: *The History of Modern Mathematics* (Rowe, D. and McCleary, J., eds.), New York, Academic Press: vol. II, 287–305.

Pelissier, A. and Tête, A., 1997, *Les textes fondateurs en sciences cognitives*, Paris, Presses Universitaires de France.

Pestre, D., 1984, *Physique et Physiciens en France, 1918-1940*, Paris, Editions des Archives Contemporaines.

Pestre, D., 1990, "La reconstruction de la physique en France après la dernière guerre mondiale", *Lettre d'information, education, recherche et industrie*, 11: 21–27.

Pickering, A., 1995, "Cyborg history and the WW II regime", *Perspectives on Science*, 3: 1–48.

Pickering, A., forthcoming, "Units of Analysis: Notes on World War II as discontinuity in the social and cyborg sciences".

Poundstone, W., 1992, *Prisoner's Dilemma. John von Neumann, Game Theory and the Puzzle of the Bomb*, New York, Anchor Books.

Rau, E., 2000, "The adoption of operations research in the United States during world war II", in: *Systems, Experts, and Computers. The Systems Approach in Management and Engineering, World War II and After* (Hughes, A. C. and Hughes Th. C., eds.), Cambridge (Mass.), MIT Press: 57–92.

Roland, A., 1985, "Science and war", *Osiris*, 2nd series, 1: 247–272.

Salomon, J.-J., 1989, *Science, guerre et paix*, Paris, Economica.

Sapolsky, H. M., 1990, *Science and the Navy: The History of the Office of Naval Research*, Princeton, Princeton University Press.

Schweber, S. S, 1986, "The empiricist temper regnant, theoretical physics in the United States, 1920-1950", *Historical Studies in the Physical and Biological Sciences*, 17: 55–98.

Schweber, S. S. and Fortun, M., 1993, "Scientists and the legacy of World War II: the case of operations research", *Social Studies of Science*, 23: 595-642.

Shapley, D., 1993, *Promise and Power. The Life and Times of Robert McNamara*, Boston, Little, Brown.

Sismondo, S., 1999, "Editor's introduction: models, simulations, and their objects", *Science in Context*, 12 (2): 247–260.

Smith, C. J., *History of the Electronic Systems Division, SAGE; Background and Origins*, AFSC Historical Publication Series, 65-30-1, t1.

Tarwater, D. (ed.), 1977, *The Bicentennial Tribute to American Mathematics, 1776–1976*, Washington, Mathematical Association of America.

Waring, S. P., 1991, *Taylorism Transformed. Scientific Management Theory since 1945*, Chapel Hill, University of North Carolina Press.

Whitfield, S., 1996, *The Culture of the Cold War*, Baltimore, Johns Hopkins University Press.

Weintraub, R. E. and Mirowski, P., 1994, "The pure and the applied: bourbakism comes to mathematical economics", *Science in Context*, 7: 245–272.

Winsberg, E., 1999, "Sanctioning models: the epistemology of simulation", *Science in Context*, 12 (2): 275–292.

Wolfowitz, J., 1971, "Abraham Wald", *Dictionary of Scientific Biography*, vol. XIV: 121–123.

Part II
Technological Knowledge and Mathematical Models in the Analysis, Planning, and Control of Modern Engineering Systems

5 Technological Concepts and Mathematical Models in the Evolution of Control Engineering

Stuart Bennett

5.1 Regulators and Servomechanisms

The fundamental concept underlying the theory and practice of control engineering is *negative feedback* (normally simply referred to as feedback). Feedback is the use of a measurement of some aspect of system behavior to correct or adjust that behavior. Artefacts exhibiting deliberate use of feedback have been extant for over two thousand years, but the English word "feedback" dates from 1920 when it was used to describe parasitic connections in a wireless amplifier which resulted in local oscillations[1].

The fundamental role of feedback control is to remove *uncertainty* from a system or to move it from one part of system whose behavior is considered to be (for some reason) critical to another part whose behavior is less important. To be able to design and implement an effective feedback control strategy, the behavior of the system in the absence of the feedback has to be described in a way that is useful to the control designer. From the designers point of view the most effective description is a mathematical model of the system which abstracts the essential features of the system behavior, a model which reduces the complexity of the behavior of the real system to a sufficient degree that effective and economic control strategies can be designed.

The presence of feedback loops in a system adds to the complexity of the system and thus to the complexity of the mathematical model necessary to predict its behavior. However, the deliberate addition of feedback can make effective control of complex systems easier (see section 2 below). Just as parasitic feedback can lead to oscillation, so can the deliberate addition of feedback and thus the concept of dynamic stability and means of determining stability are important in control engineering. The early methods of determining stability are described in section 5.4 below.

Control devices can be divided into two classes: regulators and servomechanisms. Control engineers talk about the "regulator problem" and the "servomechanism problem". Regulators are devices whose purpose is to keep some system variable – temperature say – constant, and performance is measured in terms of the ability to reject disturbances. The input of a regulator system is a constant value which represents the desired value (the set point) of the some output variable and its performance can be expressed in terms of the deviation from the set value and time to return to the set value following a disturbance. These are measures that can be obtained from simple observations and can be predicted from the solution of differential equation models derived from physical principles. Servomechanisms are devices whose purpose is to follow an input signal, they are signal processing devices, and performance is measured in terms of ability to faithfully reproduce some input signal, for example, transmission of electrical signals representing sounds over long distances as in telephone and radio systems.

Consideration of the servomechanism problem, particularly the telephone system, led to the use of input-output models, that is, mathematical descriptions of the input-output relationship which ignore the actual physical process connecting the input and output signals. The nature of the signal transmitted over telephone systems, speech, also led to a consideration of how to characterise system behavior in response to an arbitrary input signal. The appropriate performance measures to use in the servomechanisms problem are less obvious that those in the regulator problem in that they need to include some consideration of the characteristics of the input signal. As is described in section 5, modelling in terms of input-output behavior for different frequencies of the input signal proved to be fruitful and led to performance measures expressed in terms of bandwidth and gain and phase margins.

The performance of both regulators and servomechanisms can be characterised in terms of their ability to reduce *uncertainty* in the system and this led to the adoption of statistically based measures and also to the description (modeling) of input and output signals in statistical terms (see section 5).

The complexity introduced by feedback loops makes it difficult to obtain closed form analytical solutions for many real systems and this has resulted in a close relationship between control engineering and computer systems. The development of the digital computer during the 1950s made it possible to develop a new control system design approach based on using state-space models of systems (see section 7). Subsequent computer development provided the technology to support the employment of statistical techniques to build mathematical models of systems from measured data (see section 8).

5.2 Models in Control Engineering

The central concern of control engineering is the design of a control structure that will produce a *desired dynamic behavior* of a system. Implicit in the words "desired dynamic behavior" is the assumption that we are concerned with the behavior over future time from some initial state: this implies that the designer has to have available, and has to use some means of predicting the future behavior of the system. The prediction mechanism, the *model*, may form part of the control structure, or may be used solely for the purpose of providing the designer with the means of evaluating alternative control structures. In this latter role the model provides *explanation* rather than detailed prediction.

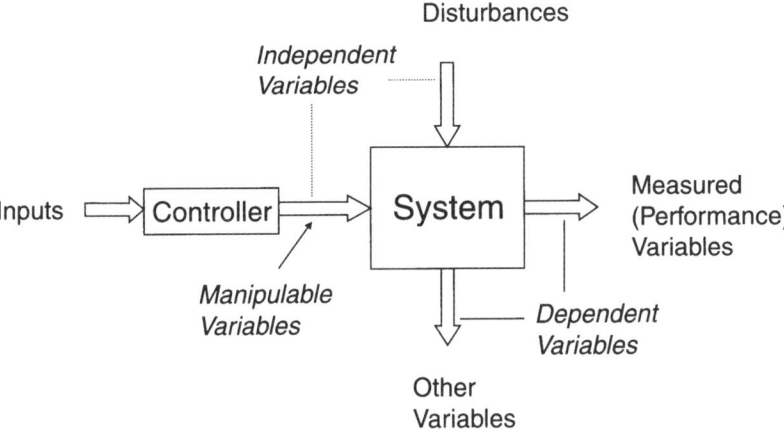

Figure 1: General control system

A *system*, in this context, can be represented as shown in fig. 1. The purpose of the controller is to adjust the controllable independent variables, the *manipulable* variables, so as to produce the desired behavior regardless of the behavior of the uncontrollable independent variables (disturbances). The desired behavior may be restricted to a small subset of the dependent variables, these are shown as *measured* (performance) variables in fig. 1 (they are also often referred to as *output* variables or *controlled* variables) and the other dependent variables are referred to as intermediate variables.

To design a controller which can force the system to produce the desired behavior it is necessary to *know* what effect the manipulable variables and the disturbance variables have on the measurable performance variables and, because future behavior of the system has to be predicted, the designer must work with a mathematical model of the system even if the system already exists in physical form.

The control engineer typically divides models into two major categories: internal models and external models. Internal models describe all the internal couplings between the variables in the system, for example, state space models. These models are built using detailed knowledge of the physical behavior of the system and they express the transfer relationships between both the internal (endogenous) variables and the external (exogenous) variables of the system. External models give the relationship between the inputs and outputs of the system. These models are obtained using the process of *system identification* and they are expressed in the form of a transfer relationship between the external (exogenous) variables of the system. This transfer relationship may be in the form of a table of data, a graph, or an analytical expression. The determination of an external model requires tests to be carried on an existing system although, for simple components, once the basic input-output model has been established it is possible to directly relate this to the physical parameters of the component.

All mathematical models, from the simple to the complex, provide prediction, however, models that provide detailed prediction of behavior, for example models used for forecasting the weather, are necessarily much more complex than those which provide a general indication of the effects of changes in the variables. Models can contain a mixture of process and data and in choosing a model for a particular purpose it is normal to choose a model structure in which the size of the combined process and the necessary data representation is the smallest that will reproduce, to some desired accuracy, the raw data. If we assume that the process is represented by a computer algorithm, then the number of bits occupied by the algorithm and the data is a measure of the size of the model. Modeling can thus be thought of reducing "the raw data to the shortest program which will reproduce the data"[2].

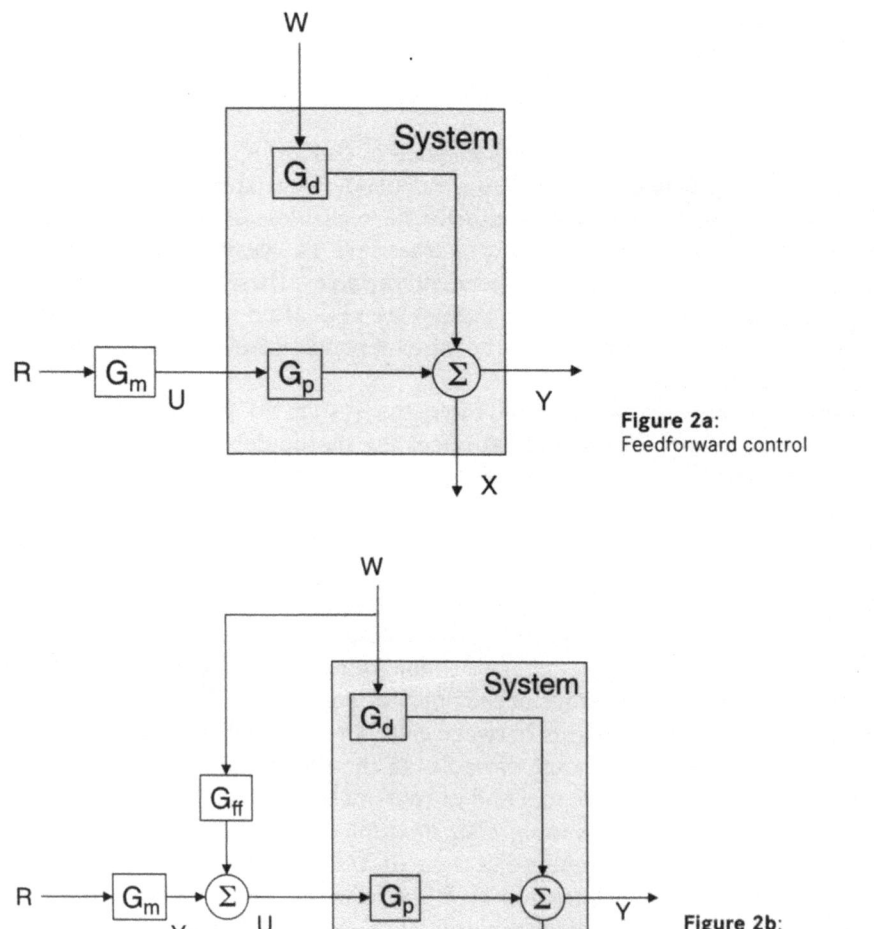

Figure 2a:
Feedforward control

Figure 2b:
Feedforward of measure disturbance

If, as shown in fig. 2a, we assume that the system to be controlled can be represented by two transfer relationships G_p and G_d, where G_p represents the relationship between the outputs Y and the manipulated variables U, and G_d represents the relationship between Y and the disturbances W. Then the controller G_m not only has to know how to manipulate U in order to get Y to follow the input command R, but it also has to predict the disturbances at time t and to modify U to counteract their effect. This implies that G_m needs to contain a model of the external world of sufficient scope and complexity to enable it to predict accurately and continuously the disturbances to which the system is susceptible.

If we assume that we can measure the disturbances, then the control structure shown in fig. 2b can be used. If we assume that R is zero, then G_{ff} has to compensate for the effects of the disturbance W and this can be done if we can make the relationship $G_{ff} = -G_p^{-1} G_d$. To do this a detailed model of the behavior of the system is needed: this is still a difficult problem but much less so than modeling the behavior of the external world.

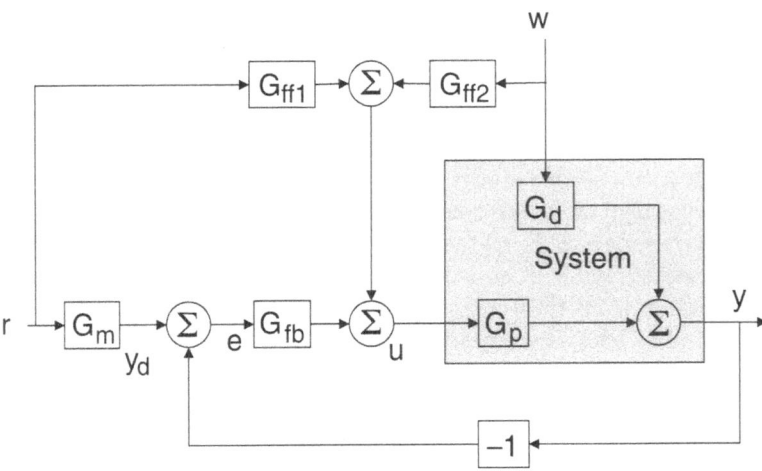

Figure 3: General control structure

A general control structure combining feedforward and feedback is shown in fig. 3. If we assume that all the sub-systems are linear and time invariant and that the transfer relationships are transformed, using the Laplace transform, into transfer functions, then we obtain for the transformed error signal $(E(s) = \mathcal{L}(e(t))$, where $e(t) = y_d(t) - y(t))$

[1]
$$E(s) = \frac{G_m - G_p G_{ff1}}{1 + G_{fb}G_p} R(s) - \left(\frac{G_p G_{ff2} + G_d}{1 + G_{fb}G_p} \right) W(s)$$

From equation [1] it follows that the transfer function $G_1(s)$ from the input command $R(s)$ and the error $E(s)$ is given by

[2]
$$G_1(s) = \frac{G_m - G_p G_{ff1}}{1 + G_{fb} G_p}$$

Thus, if G_{ff1} is chosen such that $G_m = G_p G_{ff1}$, then the error will be made zero regardless of the feedback. Alternatively if the loop gain $G_{fb} G_p$ is large then the $G_1(s)$ will be small and the error will be small.

Again from equation [1], the transfer function $G_2(s)$ relating the disturbance to the error is given by

[3]
$$G_2(s) = -\left(\frac{G_p G_{ff2} + G_d}{1 + G_{fb} G_p} \right)$$

As before, if $G_{fb} G_p$ is large the transfer function is small and hence the error is small. In this case, if G_{ff2} is chosen such that $G_p G_{ff2} = -G_d$, then the effect of the disturbance will be cancelled out.

The above shows the essential differences between feedforward control and feedback control: feedback control reduces the effect of disturbances or changes in input by dividing by a large number, whereas feedforward cancels terms. To obtain effective cancellation, the feedforward controllers must be chosen to accurately match the system, and hence feedforward controllers are sensitive to model errors. An effective feedback controller structure can be designed without using a model of the system or with a very simple model, since as shown in equation [3], providing the denominator is large the precise value is not important. The controller settings can then be determined by simple tests on the system. By using feedback, a controller can be designed which does not incorporate in its structure any explicit knowledge of the system. And until around 1940 nearly all feedback controllers – typically on-off, P, PI or PID –were designed in this way.

5.3 Model Representations

The transfer relationships (G_p etc.) can be expressed (modeled) in many different ways The most commonly used ways in control systems are described below.

Intuitive

Human beings operate with a mental model or picture of the system and use this either to solve a design problem or as part of their decision making process when directly or indirectly controlling some process. One of the most widely used controller designs, the PID (Proportional + Integral + Derivative) controller, arose from consideration of how a human being acted when steering a ship (Bennett 1984). From watching a helmsman, Nicholas Minorsky (1885-1970) realised that the helmsman had an understanding of how the ship would respond to both the movement of the helm and to the disturbances caused by winds and currents: to have such an understanding implies having a mental model of the behavior of the ship (Minorsky 1922). A problem with intuitive models is not that they are qualitative but that we do not fully model the process so as to account for all the necessary variables. A recent example of the limitations of intuitive modeling can be found in the experiments carried out by John D. Sterman which showed, using a management game, that managers typically fail to account properly for goods in the supply line (Sterman 1989).

Models derived from intuition or from qualitative observations can be formalised in a variety of ways: required time scales for dominant time constants, frequency ranges, settling times, and linguistics rules of the form "if x is *LARGE* and y is *SMALL* then *CLOSE* z". These formulations are based on the ability of human beings to summarize information into categories which have an approximate relationship with the primary data. The linguistic description *LARGE* used in the rule given above is an example this approximation. Two major bodies of theory – qualitative reasoning and fuzzy systems – have been developed to support the analysis and design of controllers using qualitative and fuzzy measurements[3]. The major difference between qualitative and fuzzy modelling is largely determined by the method used to set the quantity space. It should be noted that fuzziness does not imply randomness: the imprecision arises from the absence of any sharply defined criteria for class membership.

Analytical Models

The most commonly used models in control system design are analytical or mathematical models based on differential or partial differential equations. Using differential equations a general, linear input-output (external) model with input $u(t)$ and output $y(t)$ can be expressed as

[4] $\qquad y^{(n)}(t) + a_1 y^{(n-1)}(t) + a_2 y^{(n-2)}(t) + \cdots + a_n y(t) = b_1 u^{(n-1)}(t) + \cdots + b_n u(t)$

where the superscript n denotes differentiation with respect to time by n times. In discrete time, the model becomes

[5] $\qquad y(t_k) + a_1 y(t_{k-1}) + \cdots + a_n y(t_{k-n}) = b_1 u(t_{k-1}) + \cdots + b_n u(t_{k-n})$

where t_k are the sampling intervals. We can express analytical input-output models in the frequency domain in the form

[6] $\qquad G(j\omega) = \dfrac{y(j\omega)}{u(j\omega)} = \dfrac{b_1(j\omega)^{n-1} b_2(j\omega)^{n-2} + \cdots + b^n}{(j\omega)^n + a_1(j\omega)^{n-1} + \cdots + a_n}$

In control engineering a widely used model representation is the "transfer function representation" obtained by mapping time function relationships into complex function relationships (mapping from the time domain into the complex plane) using the Laplace transform. If $U(s) = \mathcal{L}(u(t))$ and $Y(s) = \mathcal{L}(y(t))$ then we can write

[7]
$$G(s) = \frac{Y(s)}{U(s)} = \frac{b_1 s^{n-1} + b_2 s^{n-2} + \cdots + b_n}{s^n + a_1 s^{n-1} + \ldots + a_n}$$

For internal models the so called state-space representation is commonly used. This takes the form

[8]
$$\dot{x} = Ax + Bu$$
$$y = Cx + Du$$

where x is a finite state vector represent the state of the system, u is the control input vector and y is the output vector; A, B, C, and D are matrices: A is called the state matrix, B the input matrix, C the output matrix and D is the disturbance matrix. The systems is thus modelled in terms of a set of first order differential equations which enable the trajectory of the state variables to be determined. The input and output matrices provide the information on the relationship between the states and the inputs and outputs.

Graphical Models

Plots or tables of data showing the response of the system to some known input constitute a model of a system and such plots are used for example, in the tuning of PID controllers, and can also be used to estimate the parameters of analytical models. Another widely used representation is to describe the transfer relationship in terms of a plot amplitude and phase relationships for a sinusoidal input of varying frequency, the so called frequency response. Various forms of the plot, Nyquist, Inverse Nyquist, Bode and Nichols are the basis of several control system design techniques.

5.4 Determination of Stability of a System

The desired behavior is the goal of the system and goals can be expressed in many ways: keeping one of more variables constant (the regulation problem), producing an output which follows an input (the servomechanism problem), or reaching some desired end point and at the same time satisfying some "economic" criteria (minimum fuel, minimum time). However, a fundamental requirement is that the controlled system must be *stable*[4]. A simple definition of stability for a linear system is that the outputs and all other dependent variables must have a bounded response to bounded inputs.

During the 19th century and the early part of the 20th century, a major question was, given a dynamical system, with or without explicit feedback control, how can we determine if it will be stable. The additional questions of how stable is the system and how sensitive is the system to changes in parameters began to emerge only during the 1930s. And it was only in the 1940s and 1950s that more precise performance measures based on functions of error and/or control effort, and on criteria such as minimum time or minimum fuel began to be defined and used.

The first major attempt at understanding the conditions for stability of a control system incorporating feedback was that of George Biddell Airy (1801–1892) in 1840. He studied a friction governor which was being used to control the motion of a telescope and, using classical analysis techniques, obtained a non-linear differential equation model of the system (Airy 1840). This equation could not easily be made linear and he was unable to reach any general conclusion. He returned to the problem in 1851 and concluded that the system could be stabilised by adding a device which would introduce a friction force proportional to the velocity of the oscillation (Airy 1951). The importance of Airy's work was that he drew attention to the problem of determining stability.

James Clerk Maxwell (1831–1879) in his essay "On the stability of motion of Saturn's rings" showed that the stability could be determined from the locations of roots of the an equation which modelled the dynamic behavior of the rings (Maxwell 1859). The issue of stability and the location of the roots of what we now call the characteristic equation was picked up by William Thomson (Lord Kelvin, 1824–1907) and Peter Guthrie Tait (1831–1901) in their book *A Treatise on Natural Philosophy* (1867). The characteristic equation is defined as $\det[\mathbf{A}-\Sigma\mathbf{I}] = 0$, in simple terms it is obtained by setting the denominator of the closed loop transfer function equal to zero. Thus, for the system described by equation [7], the characteristic equation is

[9] $$\Delta(s) = s^n + a_1 s^{n-1}+...+a_n = 0 \ .$$

Meanwhile Maxwell had been involved in the experiments to standardise the Ohm and through this work had become acquainted with the governor designed by Henry Charles Fleeming Jenkin (1833–1885). In a famous paper "On governors" (1868), Maxwell showed that governors working on the principle of Fleeming Jenkin, when applied to a simple machine, resulted in a third order characteristic equation $s^3 + a_1 s^2 + a_2 s + a_3 = 0$ and that for such an equation the stability of the system could be determined by simple manipulation of the coefficients of the equation (Mayr 1971a, b). For this special case the necessary and sufficient conditions for stability are: $a_1 > 0$, $a_3 > 0$, $a_1 a_2 > a_3$.

The stability problem had also been studied by Ivan A. Vyshnegradsky (1831–1895), who, in 1876, obtained a similar result to Maxwell's for a third order system. This result was shown as a design chart by setting $a_3 = 1$ and then a plot of a_1 and a_2 shows the stability region corresponding to $a_1 > 0$, $a_2 > 0$, $a_1 a_2 > 1$ (Vyshnegradsky 1876).

Maxwell attempted to find the conditions for a characteristic equation of 5th order but did not fully succeed, he was able to find two necessary conditions but was not able to show that these were sufficient. In 1874, Edward John Routh (1831–1907), using the suggestion made by William Kingdom Clifford (1845–1879) to Maxwell in 1868, was able to find the necessary and sufficient conditions for determining the stability of a 5th order equation from an examination of the coefficients (Routh 1874). He then studied the general problem in depth and in 1876 submitted an essay *A treatise on the stability of a given state of motion* for which he was awarded the Adams Prize. The essay was published unchanged in 1877 (Routh 1877). In the essay his first approach was based on the suggestion of Clifford. He then tried another approach based on a special case of Cauchy's "theorem of the argument" and Sturmian division. Using this second approach he provided a method of manipulating the coefficients of the characteristic equation so as to obtain a set of necessary and sufficient set of conditions for determining the stability of an nth order system.

The Swiss engineer Aurel Boleslav Stodola (1859–1942), in his study of the control of high pressure turbines showed that the necessary conditions for stability were that all the coefficients of the characteristic equation must be nonzero and of the same sign but he was not able to find a sufficient condition for the general equation (Stodola 1893–1894). Prompted by Stodola, a further study of the problem was done by Adolf Hurwitz (1859–1919) and the results published in 1895. Hurwitz, like Routh, used the Cauchy index theorem but, instead of using Sturmian division, he used Hermite's work on determinants (Charles Hermite 1822–1901). In 1911 Enrico Bompiani (1889–1975) showed that the Routh and Hurwitz criteria were equivalent[6].

The Routh-Hurwiz criteria, while valuable and while still used as the first check on a controller design, are limited in that they do not provide any guidance to the designer on how to modify the system either to obtain stability or to obtain a particular form of dynamic behavior.

5.5 External Models: Impulse and Frequency Response

Impulse Response

At the end of the 19th century the development of the telegraph and telephone raised a different problem to that of the governor. This problem was formulated as a signal processing problem. In fact, communication engineers were interested in the relationship between the input and output signal as it passed through a communication channel (Bennett 1993, O'Neill 1985). It was known that the signal, speech in the case of a telephone transmission, could become distorted and this raised the question as how could the transmission line be modified so as to reduce the distortion. The distortion arises because, in a dynamic system, the output signal is not instantaneously related to the input signal: the resistance, capacitance and inductance distributed along the transmission line result in a delay to the signal and this delay is not constant for all inputs. A consequence of this varying delay is that the output signal $y(t)$ depends on the behavior of the system from time $-\infty$ to time t, that is on the whole past history of the input signal and the transmission line. In modern terms, the output of a dynamic system depends on the input and the state (history) of the system.

At issue was how to model or represent the relationship between the input signal $u(t)$ and the output signal $y(t)$, given that $u(t)$ was arbitrary. The transmission line could be modelled as a lumped parameter system consisting of capacitors, resistors and inductors; thus, the dynamics could be expressed in the form of a differential equation. The difficulty was how to use this representation to determine the output signal for an arbitrary input signal (that is an infinite number of different inputs).

Oliver Heaviside (1850–1925) approached the problem by visualising himself "riding on the front of the wave", arguing that the way in which the circuit responded to the wave front could be used to characterise the circuit. By doing this he avoided the difficulty of dealing with an arbitrary signal and was able to deploy his "operational calculus", in which he replaced the differential operator d/dx by the variable p. In this way he transformed a differential equation in to an algebraic equation. Thus, for a differential equation

[10]
$$\frac{d}{dt}y(t) + ay(t) = H(t) \qquad H(t) = \begin{cases} 0 & t < 0 \\ 1 & t > 0 \end{cases}$$

(where $H(t)$ is Heavisde's unit function which he wrote as **1**), Heaviside would have written $(p + a)y = 1$ and hence

$$y = \frac{1}{p + a} \ .$$

In general, for a nth order differential equation an operator relationship of the form

$$y = \frac{1}{F(p)} H(t)$$

is obtained where $F(p)$ is a polynomial expression in p.

Equations of this type can be easily solved for a unit step input by using partial fractions to express $F(p)$ as the sum of simple terms in p and by using rules to find the equivalent time response, or by expanding $1/F(p)$ in a series of ascending powers of $1/p$ to give

[11]
$$\frac{1}{F(p)} = \frac{b_n}{p^n} + \frac{b_{n+1}}{p^{n+1}} + \dots$$

and then applying Heaviside's rule

[12] $\dfrac{1}{p^n} H(t) = \dfrac{t^n}{n!}$ obtaining the time response $y = b_n \dfrac{t^n}{n!} + b_{n+1} \dfrac{t^{n+1}}{(n+1)!} + \dots$

Heaviside did not offer any mathematical proof for the operational methods he used (they are closely related to the Laplace transform and to techniques used by

Augustin-Louis Cauchy (1789–1857) and others earlier in the 19th century), arguing instead that they gave the correct results. This led to much work by mathematicians to develop rigorous justifications for the techniques. For example, Thomas
John l'Anson Bromwich (1873–1929), in 1916, provided a mathematical justification for Heaviside's methods. However, for engineering the major impetus to the
widespread adoption and development of the transfer relationship approach came
from two sources: Vannevar Bush (1890–1974) and the work of John Renshaw
Carson (1887–1940) of the American Telegraph and Telephone Company. Bush
had become interested in solving problems relating to the stability of electricity
distribution networks and this led to develop techniques for circuit analysis which
were published in the book *Operational Circuit Analysis* (1929). He also developed several analogue computing devices to solve numerical integral equations,
the most important being the six amplifier differential analyser built in 1931[7].

John R. Carson

The impetus for Carson's work came from the attempts being made to transmit,
simultaneously, several conversations over a single pair of lines using carrier techniques. To do this the speech signal was used to modulate the amplitude of a higher
frequency carrier signal. For this method to work, cross talk between the carriers
had to be small and filters which could separate the carriers at the receiver were
needed, as well as filters which could remove the carrier from the modulated signal,
thus leaving the original speech signal. Therefore telephone system designers
needed mathematical models which would enable them to predict the effect of specific filters on amplitude modulated signals. In theory, convolution methods using
equation [13] below could be used, but such methods are difficult and laborious.

Carson wrote several papers on the topic between 1917 and 1925. In 1925 he
gave a series of lectures at the University of Pennsylvania Moore School of Engineering; the lectures were published in the Bell System Technical Journal (Carson
1925, 1926) and then as a book, *Electric Circuit Theory and Operational Calculus*
(1926). These publications give an excellent summary of his ideas and approach.
He showed that any arbitrary signal could be represented by a series of steps or
pulses separated by small time intervals and that, by using the superposition principle, the response was given by the sum of the responses to each step or pulse. As
the time interval between the steps approaches zero then the summation can be
replaced by an integral. He then proved that if $W(t)$ is the time response of a system
described by operator $G(p)$ when subjected to an unit-impulse input, the response
to an arbitrary input signal $u(t)$ is given by the convolution integral

[13]
$$y(t) = \int_{-\infty}^{t} u(t-\tau)W(\tau)\,d\tau$$

$W(t)$ is often referred to as the weighting function since each value of the input
signal is "weighted" by the response of the system to a unit impulse.

He then investigated what happens by assuming that the input signal $u(t) = A\,e^{j\omega t}$, that is a sinusoidal signal. What he then found was that he obtained transform relationships that corresponded to the Fourier transform pair. He thus obtained a frequency response relationship $y(t) = AG(j\omega)\,e^{j\omega t}$, where $G(j\omega)$, the frequency response function, is obtained from $G(p)$ simply by replacing p by $j\omega$. Carson then argued that by using Fourier's analysis any arbitrary function can be expressed as the sum of sine and cosine functions. Therefore, by taking the Fourier transform of the input and output signals, Y and U, one could write

[14] $$Y(j\omega) = G(j\omega)U(j\omega).$$

He further argued that if a signal is to be passed through two networks with frequency responses functions of $G_1(j\omega)$ and $G_2(j\omega)$, then the effect of the two networks is given by $G_3(j\omega) = G_1(j\omega)\,G_2(j\omega)$. Performing this operation, using the frequency response functions, replaces convolution by multiplication and the inverse transform can be used to obtain the time domain behavior.

There are some mathematical restrictions on the use of the Fourier transform and several people, including Harry Bateman (1882–1946) suggested using the Laplace transform; this approach gradually came to replace the use of the Fourier transform. In the Laplace transform method, the equivalent of equation [14] is $Y(s) = G(s)\,U(s)$, where $G(s)$ is known as the transfer function and $Y(s)$ and $U(s)$ are the Laplace transforms of the output and input signals respectively.

Frequency Response Models

The use of the frequency response – that is, the gain and phase relationship between input and output of a system measured for sinusoidal inputs over a range of frequencies – as a model of system behavior provides, for designers of filters, a powerful tool, since the frequency response can be found either experimentally or by calculation (if an analytical model can be obtained from direct physical analysis). The power of the technique was demonstrated by Harry Nyquist (1889–1976) in 1932. Seeking to find ways of showing that the negative feedback amplifier designed by Harold Stephen Black (1898–1983) was stable, he proved that if the frequency response locus plotted as shown in fig. 4 did not encircle the –1 point on the real axis, then the system was stable (Nyquist 1932). The plot is obtained by finding the gain and phase of the output of a system excited by a sinusoidal input signal at different frequencies and plotting these such that the gain represents the distance from the origin of the diagram and phase the angle measured in a clockwise direction from the horizontal axis. The resulting plot is actually an Argand diagram showing the real and imaginary parts of the complex frequency of the system transfer relationship $G(j\omega)$[8].

The invention of the negative feedback amplifier arose from the need to amplify telephone signals to compensate for the transmission losses over long lines. Early electronic amplifiers (circa 1920) provided the necessary amplification but introduced distortion. This distortion was acceptable when the signal passed through a

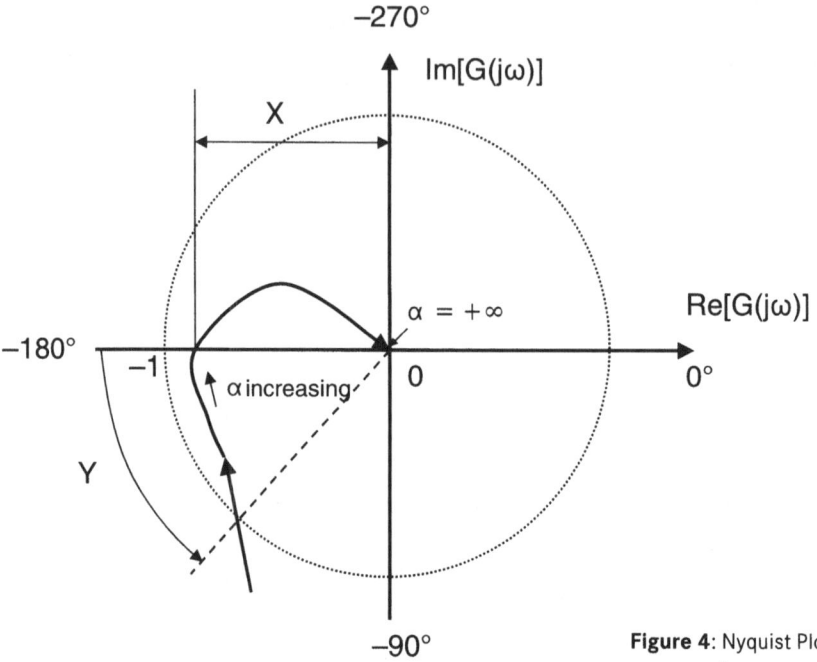

Figure 4: Nyquist Plot

few amplifiers in series but, as the number of amplifications increased, the speech the signal represented became unintelligible. The carrier systems being developed demanded higher frequency signals and the transmission loss at the higher frequencies was greater than that at the lower frequencies, hence to transmit the same distance required many more amplifiers for the carrier systems than the original single channel systems. During the mid-1920s considerable research and development work was expended on improving components and production techniques in an attempt to reducing the distortion and drift in electronic amplifiers. Black's key idea was to get the amplifier to compensate for its own shortcomings by "feeding back" into the input a proportion of the output but in a negative sense, that is if the output is +10 then a signal between 0 and –10 volts is fed back (Black 1934). Previously in communication systems "feedback" had always been positive, either to increase the gain of a low gain circuit or to create an oscillator, and many could not see how Black's amplifier could be stable (Black 1977). The concern was about the effect of delays, both in the amplifier and in the feedback circuit. Nyquist's work provided the theoretical underpinning, and it also provided a simple means of visualising the behavior by providing the designer with a graphical model of the system from which an insight into the changes necessary to modify the performance of the system could be gained.

During the mid-1930s extensive work on the frequency response method was carried out in the Bell Telephone Laboratories, particularly by Hendrik Wade Bode (1905–1982), who introduced the terms *gain margin* and *phase margin* as measures of the degree or stability of the system. The degree of stability can be interpreted as a measure of robustness of the system – Bode used the term "sensitivity" – to changes in system parameters. The distance X in fig. 4 is the reciprocal of the "gain margin", which is the amount by which the gain can be increased before the limit of stability is reached, and the angle Y is the "phase margin", which is the amount of phase shift which can be tolerated before the system becomes unstable.

Frequency response models can be expressed in several different forms and Bode introduced a method which bears his name and came to be widely used by communication and electronic engineers. The Bode plot is logarithmic plot of gain (decibels) against frequency and phase plot of phase angle against logarithmic frequency. The advantages of this plot for the designer are that it is easy to sketch on the plot approximations of the response of circuits and to quickly sketch the behavior of combinations of circuits (Bode 1940, 1945). The use of the Laplace Transform technique gradually began to replace the Heaviside operator and the Fourier transform approach outside the Bell Laboratories. Work in the early 1940s by Albert C. Hall and publication of books by Gustav Doetsch (1937), Horatio Scott Carslaw and John Conrad Jaeger (1941), Ruel V. Churchill (1944), and Murray A. Gardner and John L. Barnes (1942) resulted in the "transfer function" based on the Laplace Transform becoming the standard method of representation in control systems[9].

5.6 Stochastic and Sampled-Data Signals

The assumption underlying the treatment of feedback control design prior to the World War II was that the input signal, that is the demand on the control system, was deterministic. Inevitably measurement noise and other forms of random disturbances were present but these were assumed to be of small amplitude in the frequency band of interest. During the war several problems associated with the use of radar to track aircraft as part of anti-aircraft gun control systems and with the introduction of radar as part of the tracking system, led to a re-appraisal of this position[10].

A key element in the control of anti-aircraft guns is the prediction of the future position of the aircraft, since the time of flight of the shell, depending on the height of the aircraft, is of the order of 10 to 20 seconds. The prediction was done by a "director" (also known as a "predictor"), an analogue computer which, based on the input of the past positions of the aircraft over a period of time, attempted to predict the future position for given time ahead. This problem attracted the attention of Norbert Wiener (1894–1964) in 1940.

The predictor proposed by Wiener was a device which, on the basis of past data, built a model of the behavior of the aircraft being tracked on a least squares basis to provide the best estimate of position of the aircraft at some time t ahead of the current time. Wiener assumed that the observed tracking signal $f(t)$ belonged to a stationary stochastic (ergodic) process and the covariance Ψ of f can be found from knowledge of $f(t)$ for $t < 0$[11]. The tracking signal $f(t)$ comprises two parts:

$f_1(t)$, the true position of the aircraft (the message); and $f_2(t)$, the tracking error (the noise). The problem tackled by Wiener, with the assistance of Julian Bigelow (b. 1913), was to find some physically realisable apparatus which would give an output signal $f_h(t)$ which was, in some sense, the best estimate of the position of the aircraft, $f_1(t + h)$ at some time h ahead of the current time t. They argued that if the noise-to-noise auto-correlation ψ_{22} and the message to noise cross-correlation ψ_{12} could be found from theoretical or practical considerations, then the message to signal cross correlation is given by

[15] $$\psi_1(t) = \psi(t) + \psi_{22}(t) - \overline{\psi_{12}(t)}$$

The problem then reduces to finding, from ψ and ψ_1, a weighting function W_h which can then be used to find $f_h(t)$ using convolution since

[16] $$f_h(t) = \int_0^\infty f(t - \tau) W_h(\tau)\, d\tau$$

such that $f_h(t)$ is the best estimate in some sense of $f_1(t + h)$. Wiener used the root-mean-square error criterion and showed that $W_h(\tau)$ must satisfy the integral equation

[17] $$\psi_1(t + h) = \int_0^\infty \psi(t - \tau) W_h(\tau)\, d\tau$$

The basis of Wiener's approach was the work he had done on generalized harmonic analysis in the 1930s, when he had shown how, using an auto-correlation function, physical phenomena which exhibit apparently random behavior, but which have some underlying gross attributes, can be modeled. And, in particular, he showed that the auto-correlation function provided the link between time and frequency-response descriptions of a stochastic process (Wiener 1931).

The project was not successful, as Wiener frankly admitted in his final report stating "that an optimum mean square prediction method based on a 10 second past and with a lead of 20 seconds does not give substantial improvement over a memory-point method, nor over existing practice"[12]. The assumptions underlying the mathematical model used by Wiener assumed that the past data was available from $t = -\infty$, whereas under typical operational conditions the past data may only be available for 10 seconds and prediction 20 seconds ahead was required. Wiener and Bigelow hoped to be able to overcome this problem by using data collected from a large number of past tracks to set the values of coefficients in their system[13].

However, although the attempts to produce a practical prediction mechanism came to naught, the account of the theory, published in a restricted circulation document in 1942 and later as *The Extrapolation, Interpolation and Smoothing of Stationary Time Series* (1949), together with the book *Cybernetics or Control and Communication in the Animal and the Machine* (1948) were important in propagating feedback ideas into general systems theory, physiology and into economics.

They also had a more immediate impact in that the group in the MIT Radiation Laboratory who were involved in designing the servomechanisms for radar systems began to formulate design criteria in statistical terms, arguing that[14]

> since the input can, in general, be expressed only in statistical terms, and since the disturbances certainly can be only thus expressed, it is clear that the output oft the mechanism can be assessed only on a statistical basis. Thus what is of interest is not the exact performance of the mechanism but rather the average performance and the likely spread in performance.

During the war there were two major control problems associated with radar systems. One was the design of servomechanisms which could move the radar dish to keep it locked onto a target and the second was the integration of the radar system into the overall gun control system. A difficulty in both cases was that the tracking signal received from the radar system was intermittent, that is, the servomechanism received data at discrete moments of time evenly spaced, the so called "sampling period". In between these moments of time the servomechanism received no information. The approach to modeling this type of system was to use finite difference equations instead of differential equations. Obviously, if the sampling period is small compared to the time constants of the servomechanism, the system can be treated as a continuous system. Witold Hurewicz (1904–1956) developed an extension to the Nyquist stability criteria and a mapping to the z-plane provided a transfer function approach analogous to the Laplace transform used for continuous systems (Hurewicz 1947)[15]. Modeling in this way became increasingly important with the development and growth in use of the digital computer.

5.7 State Space Models and Optimal Control

The frequency response methods based on using computed or experimentally determined data were dominant in the field of control system design at the beginning of the 1950s. However, during the 1950s a number of interacting developments began to change the approach to modeling and to analysis and design techniques for control systems. The driving force for change was the importance which governments in both the USA and USSR attached to the control problems associated with missiles and spacecraft and the enormous resources which they made available to attack these problems. The resources supported both theoretical work but also the development of first large electronic analogue computers and, by the end of the 1950s, powerful digital computers[16].

The control problems involved the simultaneous control of a number of interacting variables, the so called multivariable problem, and the consideration of "economic" performance objectives such as minimum time and minimum fuel. The nature of the objects being controlled was such that accurate mathematical models derived from physical principles could be built and that the devices themselves could be fitted with high precision measuring instruments.

The electronic analogue computer, a development of the differential analyser designed by Bush, was essentially a set of integrators which could be interconnected in various ways so as to provide continuous time solutions to, for example, systems modelled using differential equations. When used in this way, it provides a simulation of a system. The simplest and easiest way to "program", that is, set up the interconnections for an analogue computer is to convert the nth order differential equation into a set of n first order differential equations. This is done by introducing auxiliary variables, the so called state variables. The choice of state variables is not unique and it is possible to choose variables which have physical meaning with regard to the system being simulated. By the end of the 1950s the process of modeling systems in terms of sets of first order differential equations had become widespread.

The concept of a state is important in dynamical systems theory in that it accounts for the past history of the system. The dynamical behavior of a system cannot be expressed as an instantaneous relationship between a set of input and output variables, an additional set of state variables have to be introduced to account for the past history of the system. The use of sets of first order equations in the analysis of dynamic systems has a long history with early use by François Napoléon Marie Moigno (1804–1884), but the full significance of the approach only became apparent with the work of Jules Henri Poincaré (1854–1912)[17].

The emphasis on economic performance indicators which required the achievement of some extremum value of a performance index – for example, minimum time – had a direct analogy in the variational approaches of classical mechanics found in the work of Joseph-Louis Lagrange (1763–1813) and William Hamilton (1805–1865). The classical calculus of variations techniques were advanced by Lev Semenovich Pontryagin (1908–1988) whose work has formed the basis of optimal control theory (see chapter 8). Of more immediate practical use was the work of Richard Bellman (1920–1984) who cast the dynamic optimisation under constraint problem into a form – his so called dynamic programming – in which it could be programmed for solution using the newly developing digital computer (Bellman 1954, 1957).

The optimal control formulation using the state space model provided a technique for the direct *synthesis* of controllers for a class of feedback systems, thus avoiding the time consuming design-analyse-re-design iterative method. This, together with the introduction by Rudolf Emil Kalman (b. 1930) and Richard S. Bucy (b. 1935) of the state-space equivalent of the Wiener filter (Kalman and Bucy 1961), led to the dominance of statespace methods in control theory during the 1960s and to the attempt to extend applications of the methods from aerospace problems to general industrial problems. However, exact models for industrial plant are rarely available. As early as 1961 J. J. Florentin suggested the use of a stochastic differential equation model of the form $dx = f(x,u,t)\,dt + g(x,u,t)\,dw$, where $w(t)$ was independent Wiener process, as a way to deal with plant uncertainties (Dorato 1996). However, it was the work of Howard Rosenbrock, who returned to frequency response methods, and his development of the multivariable inverse Nyquist array technique that provided a means of designing multivariable controllers for general industrial applications (Rosenbrock 1962, 1966, 1969).

5.8 System Identification

"Identification is the determination, on the basis of input and output, of a system within a specified class of systems, to which the system under test is equivalent" (Zadeh 1962)

State-space and optimal control, with the requirement for exact models generated interest in methods for estimating the parameters and states of models derived from physical principles. Parameter estimation is concerned with finding the best estimate of the coefficients (parameters) of models expressed as differential or difference equations, when the requirement is simply to have a model of the input and output relationships. State estimation is concerned with finding the best estimates of all the coefficients (parameters) of the state equations for the process. When there is some uncertainty about the number of input-output parameters required, then the problem becomes a mixed parameter-state estimator problem.

System identification is the name which has become widely accepted within the control community for the processes of parameter and state estimation (Eykhoff 1974, Ljung 1996). The techniques used rest upon statistical foundations and are also used in many other areas for the purposes of obtaining models of dynamic systems. The basic statistical methods are least squares, Markov or generalised least squares, maximum likelihood, and Bayesian.

The models obtained by parameter and state estimation techniques are often referred to as parametric models in that the coefficients of the equations have a direct connection to the physical parameters of the system being modelled. It is possible to use parameter estimation techniques to find non-parametric models. In this case, a model described by a difference equation (see equation [5]) with particular system order (m, n) is postulated, and estimates of coefficients a_i, b_i are found from observed input and output data for the process. Models obtained by these methods are often referred to as semi-physical or "grey-box".

Techniques referred to as "black-box" have also been developed in which no assumption is made regarding model structure. Neural network models are the most common form black-box model. The model is created by feeding the untrained neural network with training data obtained from the system being modeled. Once trained, the neural network then "models" the behavior of the system in that, when fed with new input data, it predicts what output the real system would produce if fed with the same data. The popularity of neural net models of systems has grown as computing power has increased since this has made possible the many thousands of iterations necessary to obtain convergence when training the net (Sontag 1993, Slotine and Sanner 1993, Ljung, Sjoberg, and Hjalmarsson 1996).

The classical parameter and state estimation techniques are based on the assumptions that there is sufficient knowledge available to select an appropriate class of model from which the "best" model can be chosen and also that the input-output (cause and effect) relationships can be determined. Also, as with black-box modeling techniques such as neural networks, it is assumed that the data set used belongs to a stochastic process (that is, the data set is not unique but is a sample from the population of time series representative of the process being modelled). Jan C. Willems has argued that often this does not represent the real situation, in that the data available is a single time series and that the lack of fit between the model and the data should not be considered as being due to the stochastic nature of the data but as indicating that the model class chosen is too simple to represent the complexity of the system. The model is thus an approximation of the system. He also argues that causality should be obtained from the data and not from a priori assumptions (Willems 1996).

The ubiquity of the digital computer has made the development and application of an extensive range of identification techniques possible and this in its turn has made the construction of accurate models, and the on-line updating of such models to deal with process drift possible. With accurate models it has become feasible to consider control structures incorporating process models, and while such systems are complex and expensive to design and implement, the potential rewards for many organisations are great. For example, in large throughput process industries – oil refineries are an example – a small improvement in controller performance of the order of 0.1% is significant if the annual value of the plant throughput is measured in billions of dollars.

5.9 Conclusion

"A process cannot be characterised by one mathematical model. A process should be represented by a hierarchy of models ranging from detailed and complex simulation models to very simple models" (Karl Astrom).

For much of the 20th century engineers and others have sought to design and build systems so as to minimise their response to expected external disturbances. For example, designers sought materials and structures whose properties were invariant over time. In doing so, they were able to reduce the effective degrees of freedom of the process and hence reduce its complexity. Thus, simple models were sufficient to support the design of both open loop and closed loop controllers. Unfortunately, a consequence of this is that the control forces necessary to move process from one operating state to another can become very large. Examples of this can be found in some early aircraft designs were large control surfaces were required and these resulted in the pilots having to apply considerable forces to the controls (Vincenti 1990). Similar problems arose in the chemical industry where plants were designed for a single operating point and it was very difficult to control them to work at a different operating point.

It is only during the period 1945 to 1950 that the full potential of utilising feedback control was recognised. The frequency response approach to controller design developed by Nyquist and Bode made it easy to combine information from *a priori* knowledge obtained from models based on physical laws with *a posteriori* knowledge obtained from measurements of process variables. This led to control engineers using both methods for the purpose of building models of systems – codified as system identification. From the 1960s onwards the growing availability of digital computers made feasible both the construction of accurate models and the deployment of such models within control schemes. Contributing to this were the significant theoretical results obtained by Kalman regarding controllability and observability (Kalman 1960) and his casting of the Wiener filtering problem into a multivariable time-response approach which resulted in the realisation of the deep and exact duality of the multivariable feedback control and multivariable filtering problems.

Central to the post 1960 developments has been the digital computer both through providing the necessary computing power for system identification in order to develop models but also for supporting the controller design process. It is only through the availability of this computing power that designs for controllers for complex multivariable systems have been able to be produced. It is also this computing power which has made on-line identification feasible and the use or predictive models within controller structures.

The success of the model-based approaches to controller design and the system identification methods for modeling in engineering and technological systems has led to attempts to apply these techniques to social, economic and political systems. While it is clear that there are feedback loops within such systems and there are manipulable variables the results from such attempts have not fulfilled the early expectations. The reasons for the lack of success are many. It is not just that the social systems are more complex than engineering and technical systems but that they also have characteristics such as long time delays which cause problems which have not been effectively solved in the simpler technological systems. There are also problems with the way in which models of such systems have been built: they are either over simplified or are based on assumptions that the data set is drawn from a stochastic process where, as in many cases, the data set is unique. In both cases the models are poor approximations of reality and hence have limited validity and predictive capacity[18]. The evolution of control engineering during the 20th century has clearly shown that quality of control achieved is dependent on the accuracy of models used to design controller and hence techniques for building models (preferably mathematical models) are necessary pre-requisite for the control of systems.

Notes

[1] Oxford English Dictionary, 2nd edition 1989. For the early history of feedback control see Mayr 1970.

[2] For detailed discussions of the issues surrounding the different types of modeling and of complexity see, for example, Simon 1996, MacFarlane 1993, Kline 1995. For a detailed theoretical account of computational complexity and its relationship to information theory see Chaitin 1987.

[3] See Leitch 1989, Weld and de Kleer 1989. Fuzzy systems were introduced into control through the work of Lofti A. Zadeh (1965, 1973) and an early application was by E. H. Mamdani (1974); a review of the early work can be found in Tong 1977.

[4] This section draws heavily on Bennett 1979 and on MacFarlane 1979.

[5] Hurwitz 1895 (English translation by H. G. Bergmann, "On the conditions under which an equation has only roots with negative real parts", in Bellman and Kalaba 1964).

[6] Hermite had published a set of necessary and sufficient conditions in 1856. Further work on the problem by A. Liénard and Henri-Albert Chipart led to four alternative sets of conditions which require the computation of about half the number of determinants as the Hurwitz conditions. See MacFarlane 1970.

[7] For background to the differential analyser see Bennett 1993, and also Owens 1986.

[8] Nyquist's original work considered a restricted class of systems; appropriate extensions to a wider class of systems were given by LeRoy A. MacColl and Albert C. Hall, and Frey K. Kupfmuller proposed a similar method of determining the stability of a system to that of Nyquist. Note that the form of the graph shown in fig. 4 is actually due to Bode.

[9] Per A. Kullstam has argued that there are advantages in using the operator approach of Heaviside (Kullstam 1991, 1992).

[10] For a general account of radar and automatic control in the war see Mindell 2000.

[11] Strictly ergodicity requires the ensemble moments and time moments to be equal. For some background discussion on ergodicity see Wiener 1948, chapter 2. For mathematical background see Masani 1990 (chapter 14) and Levinson 1966.

[12] Final Report December 1942. The memory-point method was used in the Bell Telephone Laboratories designed predictor.

[13] The assumption they made was that they were dealing with stationary (ergodic) random processes in the which the time averages and ensemble averages are identical.

[14] Ralph S. Phillips in James, Nichols, and Phillips 1947, p. 265. Chapters 6, 7 and 8 were devoted to this new approach to design.

[15] A more extensive treatment was given in Tsypkin 1956.

[16] Project Whirlwind, originally intended to support the design and building of an analogue computer, was converted into the development of a stored program digital computer. During the 1950s the government funded SAGE (Semi-Automatic Ground Environment) air defence project provided the research and development effort which resulted in many ideas and devices which assisted with the development of digital computers and provided IBM with its dominance in the market place, see for example, Edwards 1996.

[17] Poincaré, in his *Méthodes nouvelles de la mécanique céleste* (1892–1899) based his work on researches of Cauchy and Moigno (1844). See MacFarlane 1970, 1979.

[18] Early attempts to apply modeling ideas from control to economic systems were Simon 1952 and Tustin 1953. Jay Forrester's book *Industrial Dynamics* (1961) was important, as was his later book *World Dynamics* (1971). This ides in this later book underlay the Club of Rome's modeling techniques (D. H. Meadows, D. L. Meadows, Randers, and Behrens 1972). The work of the Club of Rome was severely criticised for using over simplified models and for its sensitive to small parameter variations. More recent work using data analysis methods to model complex economic and social systems avoids over simplification but there are serious questions to be asked about using techniques which are based on an assumption that the processes being modelled are stochastic.

References

Airy, G. B., 1840, "On the regulator of the clock-work for effecting uniform movement of equatoreals", *Memoirs of the Royal Astronomical Society*, 11: 249–267.

Airy, G. B., 1851, "Supplement to paper 'On the regulator of the clock-work for effecting uniform movement of equatoreals'", *Memoirs of the Royal Astronomical Society*, 20: 115–119.

Bellman, R., 1954, "The theory of dynamic programming", *Bulletin of the American Mathematical Society*, 60: 503–516.

Bellman, R., 1957, *Dynamic Programming*, Princeton (N.J.), Princeton University Press.

Bellman, R. and Kalaba, R., 1964, *Selected Papers on Mathematical Trends in Control Theory*, New York, Dover.

Bennett, S., 1979, *A History of Control Engineering 1800–1930*, Stevenage, Peter Peregrinus.

Bennett, S., 1984, "Nicolas Minorsky and the automatic steering of ships", *IEEE Control Systems*, 4: 10–15.

Bennett, S., 1993, *A History of Control Engineering 1930–1955*, Stevenage, Peter Pergrinus.

Bittanti, S. and Picci, G. (eds.), 1996, *Identification, Adaptation, Learning: The Science of Learning Models from Data*, Berlin, Springer

Black, H. S., 1934, "Stabilized feedback amplifiers", *Bell System Technical Journal*, 13: 1–18.

Black, H. S., 1977, "Inventing the negative feedback amplifier", *IEEE Spectrum*, 14: 55–60.

Bode, H. W., 1940, "Relations between attenuation and phase in feedback amplifier design", *Bell System Technical Journal*, 19: 421–454.

Bode, H. W., 1945, *Network Analysis and Feedback Amplifier Design*, Princeton (N.J.), Van Nostrand.

Bush, V. , 1929, *Operational Circuit Analysis*, New York, Wiley.

Carslaw, H. S. and Jaeger, J. C., 1941, *Operational Methods in Applied Mathematics*, Oxford, Clarendon Press (2nd ed., 1947).

Carson, J. R. 1925, "Electric circuit theory and the operational calculus", *Bell System Technical Journal*, 4: 685–761.

Carson, J. R., 1926, "Electric circuit theory and the operational calculus", *Bell System Technical Journal*, 5: 50–95, 336–384.

Carson, J. R., 1926, *Electric Circuit Theory and the Operational Calculus*, New York, McGraw-Hill.

Chaitin, G. J., 1987, *Algorithmic Information Theory*, Cambridge, Cambridge University Press.

Churchill, R. V., 1944, *Modern Operational Mathematics in Engineering*, New York, McGraw-Hill.

Doetsch, G., 1937, *Theorie und Anwendung der Laplace-Transformation*, Berlin, Springer.

Dorato, P., 1996, "Control history from 1960", in: *Proceedings of the 13th Triennial World Congress of the International Federation of Automatic Control* (Gertler, J. J., Cruz, J. B., and Peshkin, M., eds.), San Francisco, Pergamon Press: 129–134.

Edwards, P. N., 1996, *The Closed World: Computers and the Politics of Discourse in Cold War America*, Cambridge (Mass.), MIT Press.

Eykhoff, P., 1974, *System Identification: Parameter and State Estimation*, London, John Wiley & Sons.

Forrester, J. W. , 1961, *Industrial Dynamics*, Cambridge (Mass.), MIT Press.

Forrester, J. W. , 1971, *World Dynamics*, Cambridge (Mass.), Wright-Allen Press.

Gardner, M. A. and Barnes, J. L., 1942, *Transients in Linear Systems*, New York, Wiley.

Hurewicz, W., 1947, "Filters and servo systems with pulsed data", in: James, Nichols, and Phillips 1947: 231–261.

Hurwitz, A., 1895, "Über die Bedingungen, unter welchen eine Gleichung nur Wurzeln mit negativen reelen Theilen besitzt", *Mathematische Annalen*, 46: 273–280.

James, H. M, Nichols, N. B., and Phillips, R. S., 1947, *Theory of Servomechanisms*, Radiation Laboratory Series, New York, McGraw-Hill.

Kalman, R. E., 1961, "On the general theory of control systems", in: *Proceedings of Automatic and Remote Control, Proceedings of the First International Congress of the International Federation of Automatic Control, Moscow, 1960* (Coales, J. F., ed.), London, Butterworths: 481–492.

Kalman, R. E. and Bucy, R. S., 1961, "New results in linear filtering and prediction theory", *Transactions of the American Society of Mechanical Engineers. Journal of Basic Engineering*, 83, series D: 95–108.

Kline, S. J. , 1995, *Conceptual Foundations for Multidisciplinary Thinking*, Cambridge, Cambridge University Press.

Kullstam, Per A., 1991, "Heaviside's Operational Calculus: Oliver's Revenge", *IEEE Transactions on Education*, 34 (2): 155–166.

Kullstam, Per A., 1992, "Heaviside's operational calculus applied to electrical circuit problems", *IEEE Transactions on Education*, 35 (4): 266–277.

Leitch, R. R., 1989, "A review of the approaches to the qualitative modelling of complex systems", in: *Trends in Information Technology* (Linkens, D. A., ed.), Stevenage, Peter Peregrinus: 278–297.

Levinson, N., 1966, "Wiener's Life", *Bulletin of the American Mathematical Society*, 72: 27.

Ljung, L., 1996, "Developments of system identification", in: *Proceedings of the 13th Triennial World Congress of the International Federation of Automatic Control* (Gertler, J. J., Cruz, J. B., and Peshkin, M., eds.), San Francisco: 141–146.

Ljung, L., Sjoberg, J., and Hjalmarsson, H., 1996, "On neural network model structures in system identification", in: Bittanti and Picci 1996: 366–399.

MacFarlane, A. G. J., 1970, *Dynamical System Models*, London, George G. Harrap.

MacFarlane, A. G. J., 1979, "The development of frequency-response methods in automatic control", *IEEE Transactions on Automatic Control*, AC-24: 250–265.

MacFarlane, A. G. J., 1993, "Information, knowledge and control", in: Trentelman and Willems 1993: 1–28.

Mamdani, E.H., 1974, "Application of fuzzy algorithms for control of a simple plant", *Proceedings of the IEE*, 212: 1585–1588.

Masani, P. R., 1990, *Norbert Wiener 1894–1964*, Basel, Birkhäuser Verlag.

Maxwell, J. C., 1859, *On the Stability of Motion of Saturn's Rings*, Cambridge, Macmillan.

Maxwell, J. C., 1868, "On Governors", *Proceedings of the Royal Society*, 16: 270–283.

Mayr, O., 1970, *The Origins of Feedback Control*, Cambridge (Mass.), MIT Press.

Mayr, O., 1971a, "Victorian physicists and speed regulation: an encounter between science and technology", *Notes and Records of the Royal Society of London*, 26: 205–228

Mayr, O., 1971b, "James Clerk Maxwell and the origins of cybernetics", *Isis*, 62: 425–444.

Meadows, D. H., Meadows, D. L., Randers, J., and Behrens, W. W. III, 1972, *The Limits to Growth*, London, Pan Books.

Mindell, D. A., 2000, "Automation's finest hour: radar and system integration in World War II", in: *Systems, Experts, and Computers: The Systems Approach in Management and Engineering World War II and After* (Hughes A.C. and Hughes, T. P., eds.), Cambridge (Mass.), MIT Press: 27–56.

Minorsky, N., 1922, "Directional stability of automatically steered bodies", *Journal of the American Society of Naval Engineers*, 342: 280–309.

Moigno, F. L. N. M., 1840–1861, *Leçons de calcul différentiel et de calcul intégral, rédigées d'après les méthodes et les ouvrages publiés ou inédits de M. A.-L. Cauchy*, 3 vols., Paris, Bachelier.

Nyquist, H., 1932, "Regeneration theory", *Bell System Technical Journal*, 11: 126–147.

O'Neill, E. F., 1985, *A History of Engineering and Science in the Bell System: Transmission Technology* (1925–1975), AT & T Bell Laboratories.

Owens, L., 1986, "Vannevar Bush and the differential analyzer: the text and context of an early computer", *Technology and Culture*, 27: 63–95.

Poincaré, H., 1892–1899, *Méthodes nouvelles de la mécanique céleste*, 3 vols., Paris, Gauthier-Villars.

Rosenbrock, H. H., 1962, "Distinctive problems of process control", *Chemical Engineering Progress*, 58: 43–50.

Rosenbrock, H. H., 1966, "On the design of linear multivariable control systems", in: *Proceedings of Third IFAC World Congress* (McLellan, G. D. S., ed.), London, Butterworth: 1–16.

Rosenbrock, H. H., 1969, "Design of multivariable systems using the inverse Nyquist array", *Proceedings of the Institution of Electrical Engineers*, 116: 1929–1936.

Routh, E. J., 1874, "Stability of a dynamical system with two independent motions", *Proceedings of the London Mathematical Society*, 5: 92–99.

Routh, E. J., 1877, *A Treatise on the Stability of Motion*, London, Macmillan (reprinted with an introduction by A. T. Fuller as *Stability of Motion*, London, Taylor & Francis, 1976).

Simon, H. A., 1952, "On the application of servomechanism theory in the study of production control", *Econometrica*, 20: 247–268.

Simon, H. A., 1996, *The Sciences of the Artificial*, 3rd ed., Cambridge (Mass.), MIT Press.

Slotine, J. J. E. and Sanner, R. M., 1993. "Neural networks for adaptive control and recursive identification: a theoretical framework", in: Trentelman and Willems 1993: 381–436.

Sontag, E. D., 1993, "Neural networks for control", in: Trentelman and Willems 1993: 341–380.

Sterman, J. D., 1989, "Modeling managerial behavior: misperceptions of feedback in a dynamic decision making experiment", *Management Science*, 35(3): 321–339.

Stodola, A. B., 1893, 1894, "Über die Regulierung von Turbinen", *Schweizer Bauzeitung*, 22: 113–117, 121–122, 126–128, 134–135; and 23: 108–112, 115–117.

Thomson, W. and Tait, P. G., 1867, *A Treatise on Natural Philosophy*, Oxford, Oxford University Press.

Tong, R. M., 1977, "A control engineering review of fuzzy systems", *Automatica*, 13: 559–569.

Trentelman, H. L. and Willems, J. C. (eds.), 1993, *Essays on Control: Perspectives in the Theory and its Applications*, Boston (Mass.), Birkhäuser.

Tsypkin, Ya Z., 1956, "Frequency method of analysing intermittment regulating systems", in: *Frequency Response* (Oldenburger, R., ed.), New York, Macmillan: 309–341.

Tustin, A. , 1953, *Mechanism of Economic Systems*, London, Heinemann.

Vincenti, W. G., 1990, *What Engineers Know and How They Know It: Analytical Studies from Aeronautical History*, Baltimore (MD), Johns Hopkins University Press.

Vyschnegradski, I. A., 1876, "Mémoire sur la théorie génerale de régulateurs", *Comptes Rendus de l'Académie des Sciences*, 83: 318–321.

Vyschnegradski, I. A., 1877, "O regulyatorakh pryamogo deystvia" (On direct-action regulators), *Izvestiya Peterburgskogo prakticheskogo tekhnologicheskogo instituta*, 1: 21–62.

Vyschnegradski, I. A., 1877, "Über directwirkende Regulatoren", *Civilingenieur*, 22: 95–131.

Vyschnegradski, I. A., 1878, 1879, "Mémoire sur la théorie génerale de régulateurs", *Revue Universelle des Mines*, 2nd series, 4: 1–38 and 5: 192–227.

Weld, D. G. and de Kleer, J. (eds.), 1989, *Readings in Qualitative Reasoning about Physical Systems*, San Mateo (Cal.), Morgan Kaufman.

Wiener, N., 1931, "Generalized Harmonic Analysis", *Acta Mathematica*, 55: 117–258.

Wiener, N., , 1942, *The extrapolation, interpolation and smoothing of stationary time series with engineering applications*, Office for Scientific Research and Development Report 370.

Wiener, N., 1948, *Cybernetics or Control and Communication in the Animal and the Machine*, Cambridge (Mass.), MIT Press.

Wiener, N., 1949, *The Extrapolation, Interpolation and Smoothing of Stationary Time Series with Engineering Applications*, Cambridge (Mass.), MIT Press.

Willems, J. C., 1996, "From data to state model", in: Bittanti and Picci 1996: 184–245.

Zadeh, L.A., 1962, "From circuit theory to system theory", *Proceedings of the Institute of Radio Engineers*, 50: 856–865.

Zadeh, L. A., 1965, "Fuzzy sets", *Information and Control*, 8: 338–353.

Zadeh, L. A., 1973, "Outline of a new approach to the analysis of complex systems", *IEEE Transactions on Systems, Man and Cybernetics*, 3: 28–44.

6 Feedback: A Technique and a "Tool for Thought"

Antonio Lepschy and Umberto Viaro

The most significant use of feedback is undoubtedly related to automatic control. Its first technological applications, often unconscious, are very old and well documented. Here we develop some considerations on the reasons why the inventors of the oldest control systems might not have been fully aware of the feedback nature of their devices (for which feedback models have only recently been proposed to explain their operation) and the reasons why, instead, feedback has consciously been employed in the technical applications of electronics since the first decades of the 20th century (to modify the performance of amplifiers and oscillators); it is worth noting that the word *feedback* has been coined in those years. Furthermore, feedback is also effectively exploited in mathematics, in two distinct, yet related, fields: the algorithms whose flow-chart includes loops and the analog computation for solving ordinary differential equations. Finally, feedback is the key element of many explanatory models for phenomena studied by the physical and biological sciences, as well as by the sciences of man and society; the latter models often exhibit an intrinsic interest from a philosophical viewpoint too.

6.1 Basic Elements of a Feedback Control System

The simplest representation of a feedback system is the one shown in fig. 1, where: P denotes the plant in which the process to be controlled takes place; C denotes the controller; and T denotes the transducer, if any, by which the variable to be controlled is transformed into a variable whose value is related (in a suitable scale) to that of the controlled variable and whose physical nature makes it comparable to the signal acting as the reference value (or course) to be reproduced by the output of the process.

An alternative representation, perhaps more useful for the present purposes, is shown in fig. 2, in which: block P indicates again the plant to be controlled; block

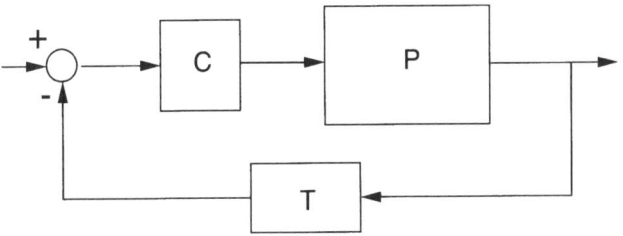

Figure 1:
Simple representation of a feedback control system

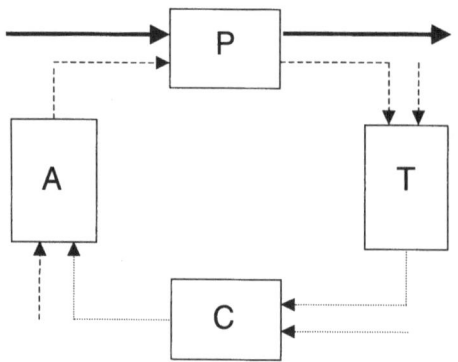

Figure 2:
Alternative representation of a feedback control system

T represents the subsystem by which the information on how the process behaves is obtained from the plant (and from its environment); block *C* represents the subsystem that processes the information on the actual plant behavior and on the objectives to be attained; and block *A* represents the actuating and amplifying subsystem by which, on the basis of the information processed by block *C*, the control action is exerted on the plant.

The input to subsystem *P* consists not only of a flow of matter and/or energy, denoted by a solid line, but also of an action (at a not negligible power level), denoted by a dashed line, which modulates the flow entering the plant. The main output of *P* is also a flow of matter or energy, again denoted by a solid line.

The input to subsystem *T* is formed by physical quantities (at a low power level) obtained from the plant and, possibly, its environment; they are denoted by a dashed line. The output of *T*, denoted by a dotted line, is the information on these quantities, to be supplied to subsystem *C*.

Subsystem *C* receives, besides the above information, another information input, denoted by a dotted line too, concerning the objectives aimed at. The output of *C* is the information obtained by processing its inputs and used to control subsystem *P* through subsystem *A*.

Finally, the input of subsystem *A* is the information processed by *C* and, if necessary, the energy required to exert the proper action on *P*, which is denoted by a dashed line. Note that an external energy source might not be necessary since, sometimes, the energy associated with the physical quantity carrying the information processed by *C* could be enough to this purpose. The functions of subsystem *A* can be of two types: (a) amplification of the power level of the ouput of *C* to the power level necessary to influence *P*; (b) actuation, i.e., generation of the physical variable acting on *P* (e.g., a mechanical torque) from the amplifier output (e.g., an electric voltage, in which case the actuator is an electric motor).

In conclusion, fig. 2 points out that the basic functions are performed by a subsystem *P* that manipulates a flow of matter or energy, and by a subsystem *C* that processes the information flow; subsystems *T* and *A* may be regarded as the interfaces linking these two primary subsystems.

6.2 Feedback Models of Some Technological Systems: Were their Inventors Aware of Such a Structure?

When the four subsystems of fig. 2 are physically distinct, the presence of feedback can easily be detected; when, instead, a unique device performs more than one of the functions represented in fig. 2, the identification of feedback becomes more problematic. The second situation tipically occurs in the earliest control devices (that have been interpreted by means of a feedback model only recently). A lot of automatic systems date back to the Hellenistic period, especially to scientists and technicians operating in Alexandria at the times of the Ptolemies. Interesting improvements of some of these devices have been realized in the Byzantine and Islamic world.

Basically, they are hydraulic devices for level and flow control, described and interpreted by means of feedback models in the classical book by Otto Mayr (1970). Here, we shall consider a typical example, namely the hydraulic clock of Ktesibios (fl. 270 B.C.), which is reconstructed in fig. 3 and interpreted in fig. 4 through a block diagram with two feedback loops. A unique element, i.e., the float valve F, performs the tasks of subsystems T, C and A of fig. 2. In fact, it transduces the level of the first vessel (task of block T) but, at the same time, manipulates this information by comparing it with the level reference (task of block C) and, finally,

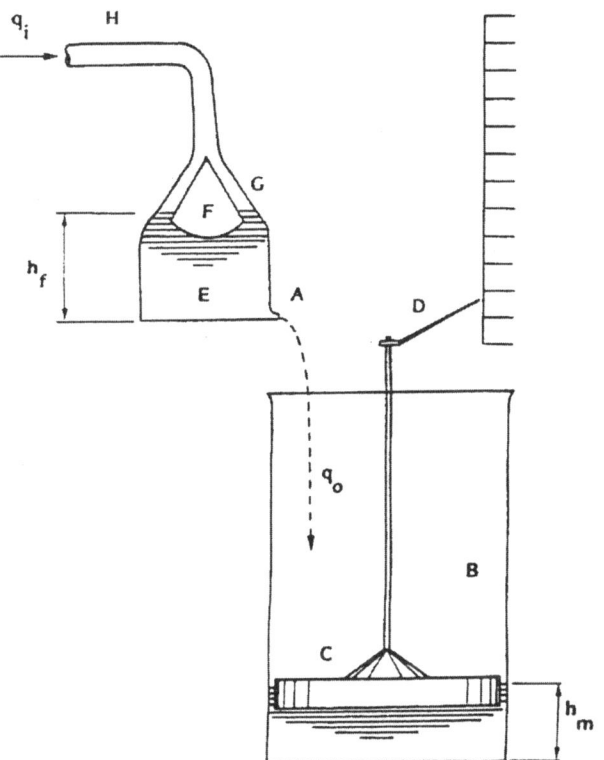

Figure 3:
Schematic reconstruction of Ktesibios' clock

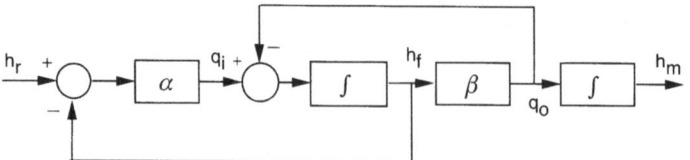

Figure 4: Block diagram describing the operation of Ktesibios' clock: h_r: reference level for the conical float valve F in the primary vessel E; h_f: actual level in the primary vessel (dependent on the integral of the difference between inflow q_i and outflow q_o); h_m: level in the secondary vessel B, connected with the indicating element D (dependent on the integral of q_o); q_i: flow from the reservoir to the primary vessel (dependent on the position of F, i.e., on $h_r - h_f$); q_o = flow from vessel E to vessel B (dependent on h_f)

on the basis of this comparison modifies the flow entering the first vessel (task of block A). There is no need for an external power supply (i.e., a power amplifier) in this device, because the buoyancy itself provides, whenever necessary, the force required to close the valve. These remarks explain why, in our opinion, Ktesibios was not aware of the feedback nature of his contrivance.

Similar considerations also hold for a substantially different timepiece, the medieval mechanical verge-and-foliot clock (Lepschy, Mian, and Viaro 1992). Fig. 5 portrays the essential parts of such clock, in which the average steady-state velocity of the wheel moved by a weight is kept constant. Without the verge-and-

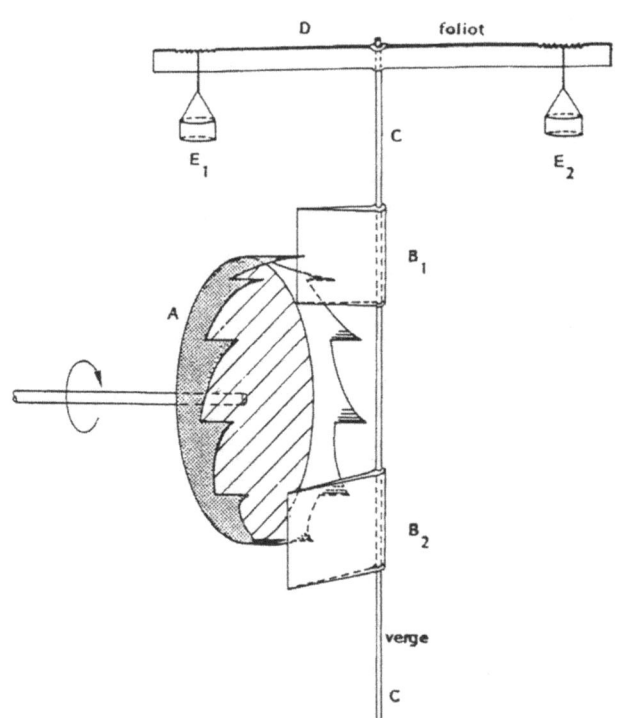

Figure 5: Schematic representation of the verge-and-foliot escapement:
A : crown wheel; B_1 : upper blade; B_2 : lower blade; C : verge; D : foliot, whose moment of inertia may be varied by shifting weights E_i (the torque applied to the horizontal shaft connected with the crown wheel is supplied by weights or by a spring)

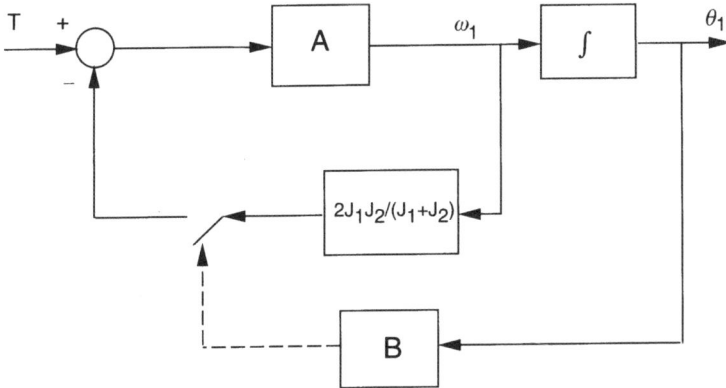

Figure 6: Block diagram describing the crown-wheel motion of the verge-and-foliot clock in the case of inelastic collisions; the transfer function of block A is $1/[(J_1 + J_2)s + F_1 + F_2]$; block B actuates the sampler; ω_1 and θ_1 are the angular speed and displacement of the crown wheel (for a detailed explanation of the block diagram and the meaning of coefficients J_i and F_i, see Lepschy, Mian, and Viaro 1992)

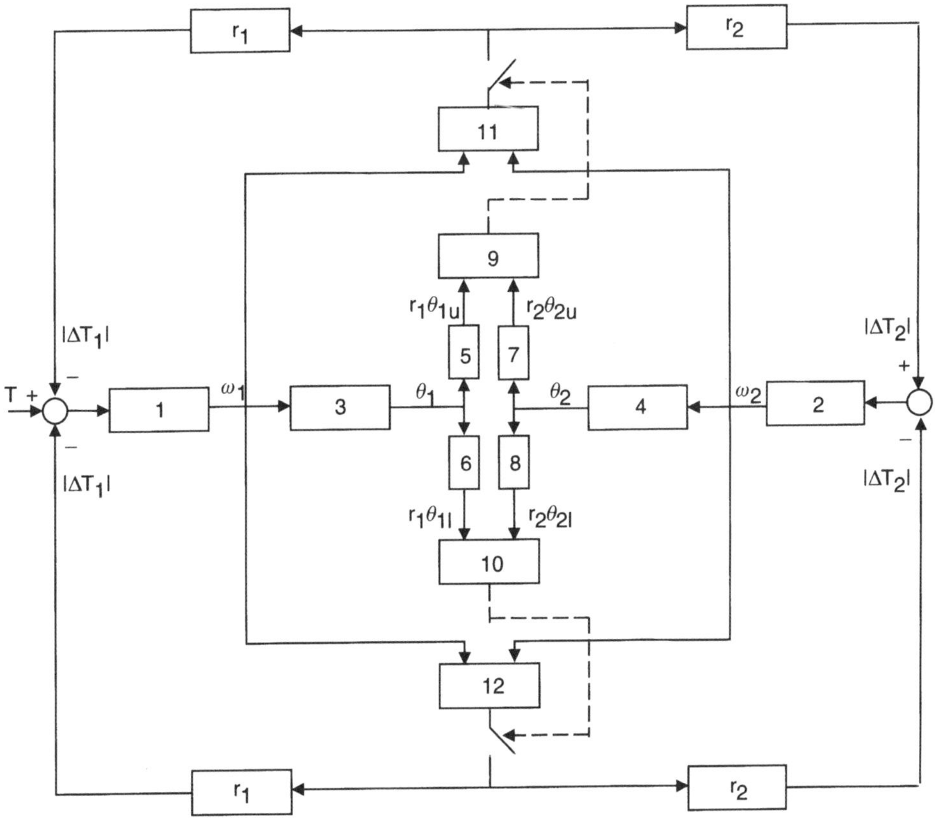

Figure 7: Block diagram describing the crown-wheel motion of the verge-and-foliot clock in the case of elastic collisions (Lepschy, Mian, and Viaro 1992)

foliot escapement its speed would increase; its average turns out to be constant because of the periodic collisions between the crown wheel and the two blades B_1 and B_2 whose planes form a suitable angle. Fig. 6 shows the related block diagram for the case of inelastic collisions, whereas fig. 7 refers to the case of elastic collisions. The presence of feedback loops in both diagrams explains why the clock angular velocity remains constant.

In this case too, there is not an immediate relation between the physical structure of the device (fig. 5) and the topology of the block diagrams describing its operation (figs. 6 and 7). Therefore, we can safely conclude that the technicians who invented and improved the above clocks were not conscious of the presence of feedback and of the reasons why such sophisticated feedback systems ensure the regularity of the clock motion. Note, also, that in this device, as well as in that of Ktesibios, the reference input is not represented by a physical quantity.

It has been stated that the first genuine feedback control system conceived in modern Europe, independently of ancient models, is the temperature regulator of Cornelis Drebbel (1572–1633) used in his thermostats for an incubator and an athanor (alchemic furnace). The remarkable progress in the control field that took place in Europe in the 17th century has a precise cause: temperature control could not be conceived before the invention of devices suitable for its measurement, that is, thermometers, acting as transducers of the controlled variable into another variable. Indeed, Drebbel in his thermostats makes use of rudimentary thermometers to move the register of the furnace smoke flue. In this case too, however, the identification of the feedback structure is not straightforward and, contrary to Mayr's opinion, the present authors are inclined to think that Drebbel was not fully conscious of it. Such conjecture is also based on the fact that, to control temperature, the manipulated variable is the smoke flow, i.e., a quantity placed downstream with respect to the heat source.

Another instance of feedback control in that period is the pressure regulator of Denis Papin's (1647–ca. 1712) cooker, in which the reference signal is supplied by the weight linked to the valve and the feedback variable is the force exerted by the pressure on the valve plate. The situation is similar to that of Ktesibios' clock, where the valve simultaneously acts as transducer (block T in fig. 2), comparator (block C) and actuator (block A).

James Watt's (1736–1819) steam-engine speed regulator uses the centrifugal flyball governor as velocity-displacement transducer and the lever connected with it as comparator and actuator. Today it is rather easy to build a feedback block diagram to describe the operation of this device, but it is not likely that Watt's ideas in this regard agreed with ours; a hint may be given by the use of the term *governor* to denote the transducer only.

It is a common opinion that the well-known paper "On governors" by James Clerk Maxwell (1831–1879) is the first contribution to the mathematical treatment of control problems (Maxwell 1867–1868). It begins with a synthetic and efficacious definition: "A Governor is a part of a machine by means of which the velocity of the machine is kept nearly uniform, notwithstanding variations in the driving power or the resistance" (p. 270). This definition may be considered reductive, because it is limited to the regulation of velocity (and not, for instance, to the regulation of pressure or temperature) and to the case in which the set-point is constant.

Maxwell takes into consideration various regulators in use at his time, from that of Watt, to those of James Thomson (1822–1892), Henry Charles Fleeming Jenkin (1833–1885), William Thomson (Lord Kelvin, 1824–1907), Jean Bernard Léon Foucault (1819–1868) and, eventually, to those based on Carl Wilhelm Siemens' (1823–1883) principle and employed by the Astronomer Royal George Biddell Airy (1801–1892) for the chronograph and equatorial of Greenwich Observatory. His attention, however, concentrates on the mathematical treatment of the regulation problem rather than on its technological aspects: "I propose at present, without entering into any details of mechanism, to direct the attention of engineers and mathematicians to the dynamical theory of such governor"(p. 271). Therefore Maxwell addresses his study particularly to engineers, who may take advantage of the theory to design regulators, and to mathematicians, whose task will be to solve the problem left open in the general case.

In fact, Maxwell correctly formulates the differential equations describing the control system behavior in terms of the relevant parameters and recognizes that what he calls "the disturbance", that is, the deviation from steady-state, includes addenda which may "continually increase, continually diminish, be an oscillation of continually increasing amplitude, be an oscillation of continually decreasing amplitude" and states that the condition preventing the occurrence of the first and third situations "is mathematically equivalent to the condition that all possible roots, and the possible parts of the impossible roots, of a certain equation shall be negative" (Maxwell calls the real roots of an algebraic equation "possible roots", its complex roots "impossible roots" and the real parts of the complex roots "possible parts of the impossible roots"). In this regard Maxwell says: "I have not been able completely to determine these conditions for equations of a higher degree than the third; but I hope that the subject will obtain the attention of mathematicians" (ibid.). Indeed, the invitation was accepted by Edward John Routh (1831–1907), who provided the first systematic solution to the problem shortly afterwards (Routh 1877; on the problem of stability, see chapter 5).

Now, we may wonder whether Maxwell was explicitly aware of the presence of feedback in his control systems. In some respect we are induced to think that he was, in some other that he was not. For instance, with reference to the force responsible for the braking action he writes: "When the velocity increases, this force increases, and either increases the pressure of the piece against a surface or moves the piece, and so acts on a break or valve"; this sentence favours the conjecture that he was conscious that an increase in velocity produces the increase of quantities related to the braking action, but this is not expressed in explicit terms. The following Maxwell's statement might suggest that he was considering the phenomenon from a different perspective: "an increase of the driving-power produces an increase of velocity, though a much smaller increase than would be produced without the moderator". In fact, in the devices called by Siemens (and by Maxwell) "moderators", feedback is a characteristic of the explanatory model today used for describing their behavior rather than a structural feature of the physical system. Again, a different impression is conveyed by the sentence: "But if the part acted on by centrifugal force, instead of acting directly on the machine, sets in motion a contrivance which continuously increases the resistance as long as the velocity is above its normal value, and reverses its action when the velocity is below that value, the governor will bring the velocity to the same normal value whatever variation (within the working limits of the machine) be made in the driving point or the resistance" (ibid.). It would seem, however, that such intuition of feedback is limited to the systems called astatic today.

This understanding of feedback is instead clearly present in two papers presented in 1876 to the Royal Society by William Thomson on the solution of differential equations by means of mechanical integrators (Thomson 1876a, b; see section 6.4), in which it is explicitly stated: "So far I had gone and was satisfied, feeling I had done what I wished to do for many years. But then came a pleasing surprise. Compel agreement between the function fed into the double machine and that given out by it. This is to be done by establishing a connexion which shall cause the motion of the centre of the globe of the first integrator of the double machine to be the same as that of the surface of the second integrator's cylinder" (Thomson 1876a: 270). Here, however, we are not dealing with control problems, rather with the exploitation of feedback for computational purposes.

Feedback has been consciously applied to design electronic amplifiers and oscillators in the first decades of the 20th century. In the thermionic valve, especially the triode, the voltage applied to the grid is immediately interpreted as an input signal and the voltage applied to the load, connected to the plate, as the amplifier output signal. If the anodic voltage is applied to the grid through a suitable filter, the presence of a feedback loop becomes evident. This explains why the word feedback was first used in 1923 in the journal *Wireless* devoted to electronics and telecommunications. It is also significant that the contributions of Harold Stephen Black (1898–1983), Hendrik Wade Bode (1905–1982) and Harry Nyquist (1889–1976) to feedback system theory referred to the study of feedback amplifiers, whereas such contributions have only later been applied to control problems. This happened during the World War II in the military field (to point aerials and weapons). The applications of the above methods in an industrial context dates from the post-war period.

6.3 Feedback Loops in Mathematics and Computer Science

An important feature of the use of feedback in electronic amplifiers consists in substantially modifying the response characteristics (like gain and pass-band) of the devices operating autonomously. A similar effect arises in the field of numerical algorithms when a loop is inserted in a flow chart. To show this, let us assume that an arithmetical unit capable of executing only additions, subtractions and multiplications is available. A (finite) algorithm using such a unit according to a flow-chart without loops can only compute algebraic sums, products of sums and sums of products; it cannot perform divisions, which instead can be obtained from the same unit by adopting a flow chart with a loop. Specifically, the flow chart of fig. 8 allows us to easily compute the inverse x_{n+1} of a given number A with an error lower than $1/M$ starting from an initial guess x_0. By multiplying the dividend by the inverse of the divisor, the quotient is then obtained without upgrading the arithmetical unit.

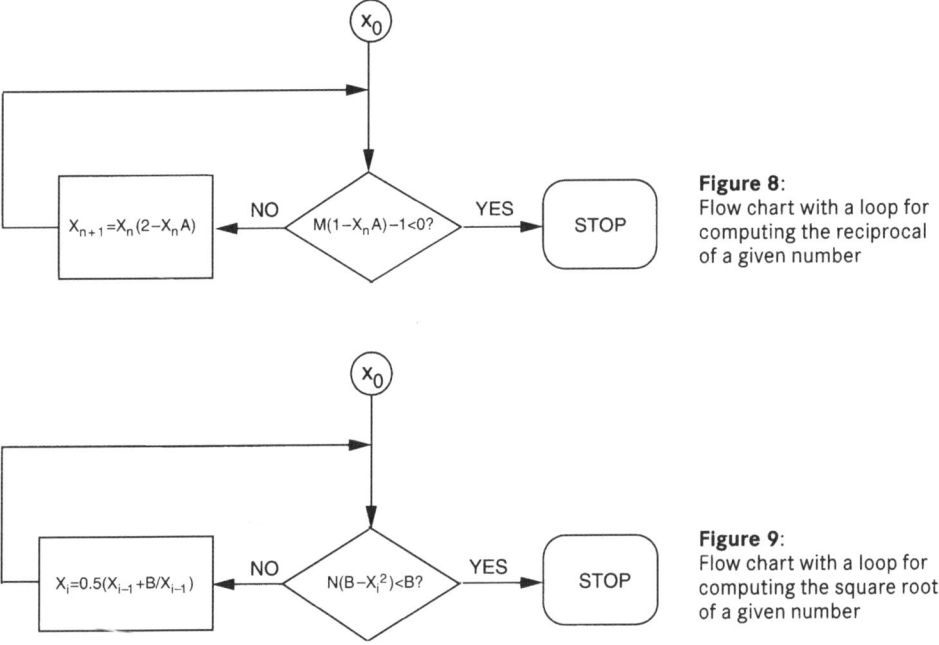

Figure 8:
Flow chart with a loop for computing the reciprocal of a given number

Figure 9:
Flow chart with a loop for computing the square root of a given number

With an analogous procedure, the square root x_{n+1} of a number B is obtained with an error lower than $1/N$ starting from an initial guess x_0, as shown in fig. 9. A general approach to this kind of procedures can be found in an often quoted paper by Dana Scott (1971).

Another important use of feedback in computation is related to the solution of ordinary differential equations by means of analog computers (today obsolete and replaced by programs, such as Simulink®, reproducing the layout of the analog computer). This idea was first exploited by William Thomson in his two 1876 papers on the subject; in particular, the first of them well documents the psychological path to arrive at a feedback diagram, as shown in the previous section. The problem he considered was to find a solution $u(x)$ of the differential equation:

[1]
$$\frac{d}{dx}\left[\frac{1}{P(x)} \cdot \frac{du}{dx}\right] = u(x)$$

in which $P(x)$ is a known function of the independent variable x. Equation [1] was given the form:

[2]
$$u(x) = \int_0^x P(\xi)\left[C + \int_0^\xi u(\zeta)\,d\zeta\right]d\xi$$

where C is a constant of integration. Of course, equation [2] does not directly supply the solution of [1] because function u appears on both sides, but its structure suggested to Thomson the idea of using the configuration of fig. 10 with two mechanical integrators I_a and I_b to obtain a better approximation $u_1(x)$ from a first

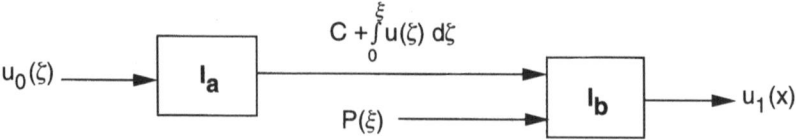

Figure 10: Cascade connection of two mechanical integrators I_a and I_b

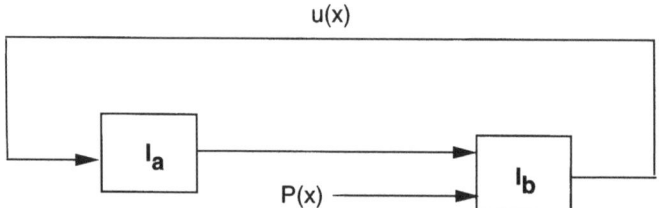

Figure 11: Feedback system obtained from the open-loop diagram of fig. 10 by connecting the output of integrator I_b with the input of integrator I_a

guess $u_0(x)$. The computation was then repeated by replacing $u_{i-1}(x)$ with $u_i(x)$, until the output of an iteration was sufficiently close to its input. This led Thomson to the conclusion of "forcing" the coincidence between the two functions by connecting the output of I_b with the input of I_a, according to the feedback diagram of fig. 11.

The above examples fit well into the general frame of so-called problem solving, which falls into the realm of artificial intelligence. This does not seem to be a suitable place for a formal treatment; so, we shall limit the discussion to the main concepts. Solving a problem can be regarded as a mapping from the set of data to the set of solutions. To obtain a solution more easily, a problem can be either reformulated or decomposed into simpler subproblems. The most trivial case is the one in which the original data are taken as the data for a simpler subproblem whose solutions form the data of a second subproblem; this, in turn, admits as its solutions those of the original problem. The corresponding graph does not exhibit loops.

However, it may happen that the data of a subproblem A consist not only of those of the original problem but also of others, not known *a priori*, which are the solutions of another subproblem B whose data are, in turn, solutions of subproblem A, so that the procedure seems to be at a standstill. Nevertheless, provided some convergence conditions are satisfied, it is indeed possible to solve the composite problem $A + B$ as follows: get started with a guess of the unknown data of A; compute the corresponding solutions of A; take these solutions as the data for B; compute the corresponding solutions of B; replace the initial guesses by these solutions; and then repeat the procedure until a suitable norm of the difference between the solutions of two consecutive iterations is sufficiently small.

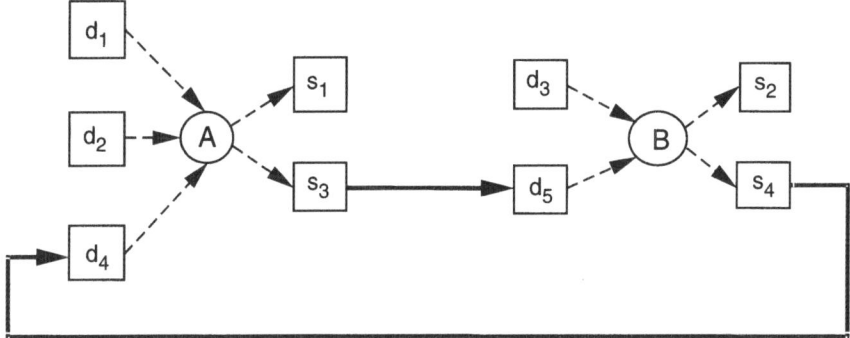

Figure 12: The missing datum d_4 (needed for the solution of problem A) is supplied by the solution s_4 of problem B

It is easily realized that the mentioned procedures for computing the inverse or the square root of a number and for solving equation [1] fit perfectly into the above general picture. The iterations correspond to a feedback loop that can be given the form of fig. 12.

6.4 The Role of Feedback in Explanatory Models

The interest of feedback is not limited to its use for the design of certain *artificial* systems in technological applications or for the solution of mathematical problems, because it also supplies an attractive model for interpreting many phenomena of different kinds. In this regard it seems useful to make some preliminary considerations. Mathematical models (or more generally, non-physical models including those conceptual models where the formalization phase is at its initial stage and the quantification phase has yet to be approached) are commonly subdivided into classes at whose extremes we may place *explanatory* models and *predictive* models. The first ones correspond to a formalization of a hypothesis; their interest consists more in the elegance with which this hypothesis is translated into formal relations than in their ability to account in a reliable way for the relevant phenomena. A predictive model, instead, may be considered effective if it provides reliable forecasts. In such a context it often happens that the main advantage of an explanatory model consists in its *simplicity*, while almost always a predictive model must be rather complex and articulated.

In the next sections, we consider explanatory models that are deliberately simple. Therefore, the reader must not expect from them a detailed, realistic and thorough description of the phenomena, as this would imply a greater model complexity. Nevertheless, we feel that the following examples have an intrinsic interest since they allow us to verify how feedback systems, in spite of the simplicity of their component parts, may exhibit behavior different from that of the interacting subsystems.

Some Explanatory Feedback Models in the Field of Physics

Speaking about explanatory feedback models in the field of physics may be ambiguous since it is often arbitrary to distinguish the situations in which feedback is an intrinsic characteristic of the system from those in which feedback arises from our interpretation.

Let us refer, for example, to Newton's second law of motion, usually expressed in the form:

[3] $F = ma$

which may be phrased as: *The force F applied to a body of mass m makes it accelerate with an acceleration a proportional to the force itself.* In this case, we recognize a forward path that leads from F to a and do not see any feedback path from a to F. Yet, according to d'Alembert's principle, this law of dynamics can be presented as a law of statics, expressing the balance of two opposing forces:

[4] $F - ma = 0$

which corresponds to the feedback diagram of fig. 13. Here the forward path accounts for the fact that a difference, if any, between the action F and the reaction ma would give rise to a change of a such that it annihilates this difference, whereas the forward path consists of an element with infinite gain.

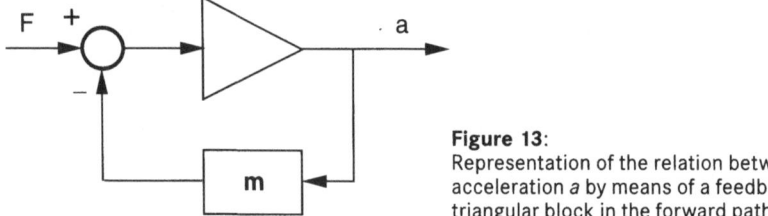

Figure 13:
Representation of the relation between force F and acceleration a by means of a feedback model (the triangular block in the forward path has infinite gain)

Feedback diagrams of the same kind can be adopted every time Newton's third law of motion applies: "To every action there is always opposed an equal reaction: or, the mutual actions of two bodies upon each other are always equal, and directed to contrary parts" (*Principia*, English translation by Andrew Motte, 1729). This principle of action and reaction has its counterparts in other fields of physics and chemistry. A particular case, very familiar to people involved in the study of electromagnetic phenomena, is the "rule" due to Heinrich Friedrich Emil Lenz (1804–1865), usually stated in the following terms: "A current brought into action by an induced electromotive force always produces effects which oppose the inducing action". The rule of Lenz is represented by the block diagram of fig. 14, where Φ_e is the external flux, Φ_f is the feedback flux (generated according to the mentioned rule) and $\Phi = \Phi_e - \Phi_f$; the time derivative of this difference produces the e.m.f. e; this determines the current i, which in turn generates flux Φ_f.

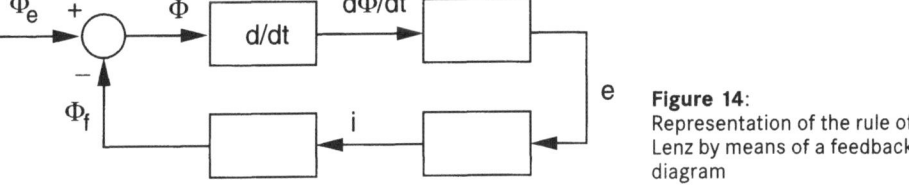

Figure 14:
Representation of the rule of Lenz by means of a feedback diagram

An almost identical statement corresponds to the "principle" of physical chemistry due to Henry Louis Le Chatelier (1850–1936): "In a system at equilibrium, the change of a parameter causes a transformation which, if it took place independently, would produce a change of the same parameter in the opposite direction". All these principles state that the system develops forces which counteract the disturbances and restore a state of equilibrium; they can be derived from the principle of minimum effect.

Some Feedback Models in the Life Sciences

More suggestive and interesting are perhaps the feedback models used in biology, in particular those accounting for homeostasis. The term "homeostasis" was introduced by Walter B. Cannon (1871–1945) in his book *The Wisdom of the Body* (1932): such wisdom consists in the aptitude for keeping the internal environment unchanged, despite the variations of the external environment. A typical instance of homeostasis is the regulation of the internal temperature in warm-blooded animals (37 °C in human beings). The result is obtained by means of numerous cooperating feedback mechanisms. Therefore, it is not very meaningful to consider only one of them; on the other hand, the model would become too complicated even if the analysis were limited to the most important ones.

A meaningful example is provided, instead, by the simplified glycemia control mechanism shown in fig. 15, where: g indicates the actual glucose concentration (glycemia), g^* represents the information concerning the desired level of g, i represents the actual insuline concentration in the blood, i^* its reference value, u_g denotes the exogenous glucose flow which is added to the endogenous flow provided by controller C, whereas the controlled process is represented by the subdiagram denoted by P. This simply performs the integration of the difference between glucose intake and consumption (proportional to g according to coefficient K_{gg}). Clearly, a more realistic description would require a much more complicated model including many interacting feedback loops.

Population dynamics and ecology are other fields in which feedback models may be used for explanatory purposes and, sometimes, for predictive purposes as well. The simplest model to represent the dynamics of a population in the absence of immigration and emigration is shown in fig. 16 a) and b). Diagram a) uses two distinct feedback channels: the birth flow gives rise to a positive feedback and the death flow to a negative feedback. Diagram b) combines the two feedback channels into a single channel, corresponding to a positive feedback if $k_b > k_d$

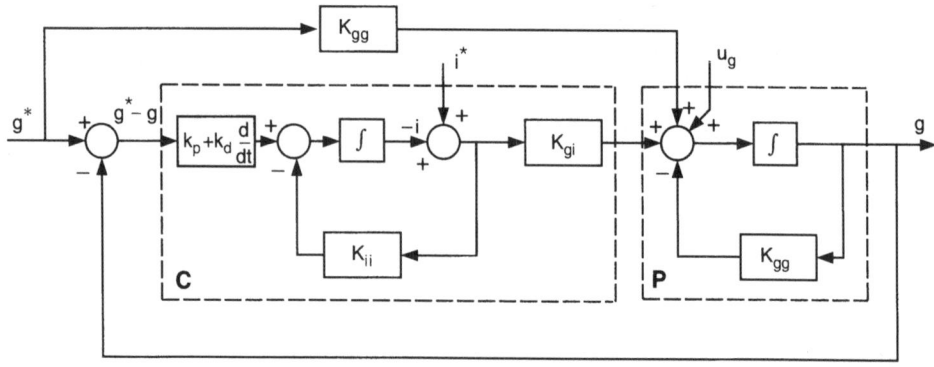

Figure 15: Multiloop block diagram representing the glycemia control mechanism

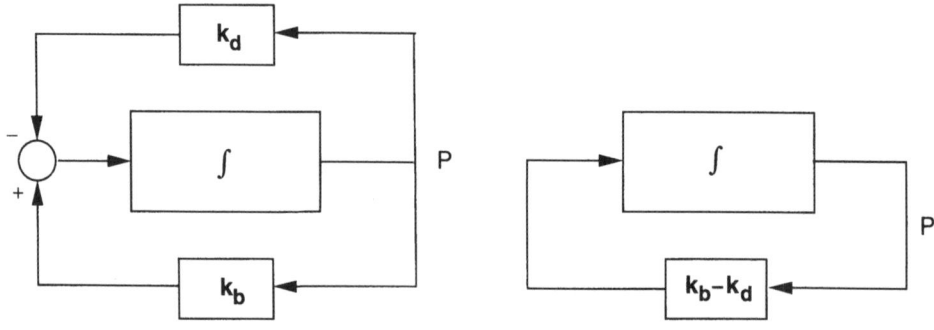

Figure 16: Block diagrams accounting for the growth of a population (without immigration and emigration)

and a negative feedback if $k_b < k_d$: in the first case the population increases with constant doubling time, in the second it decreases with constant halving time (provided the difference between k_b and k_d is constant).

A more accurate model is represented in fig. 17, where the entire population is subdivided into three age-groups corresponding to the pre-fertility, fertility and post-fertility ages.

The above representations are adequate if the population needs are very far from the carrier capacity r, i.e., from the flow of resources actually available: the population level that can be sustained by the carrier capacity is denoted by P_r. When population P approaches P_r, the rate of change dP/dt is not related to P through a constant coefficient, but via a factor depending on $P_r - P$; in the simplest model we have:

[5]
$$\frac{dP}{dt} = k(P_r - P)P$$

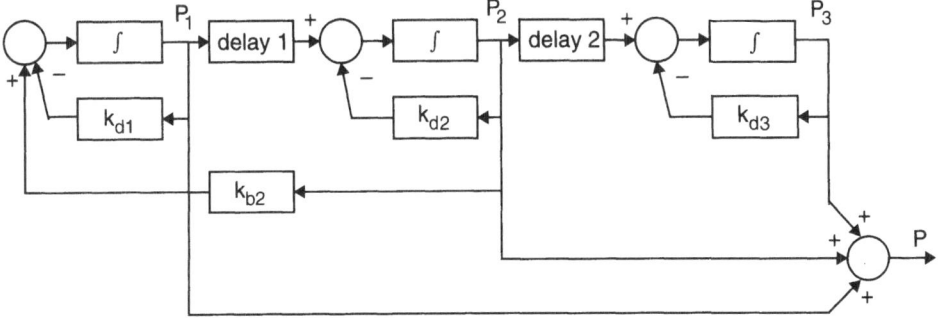

Figure 17: Block diagram accounting for the growth of a population subdivided into three age-groups

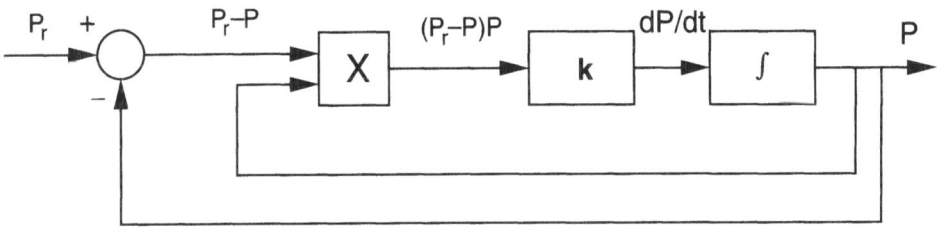

Figure 18: Block diagram representing the growth of a population with carrier capacity r

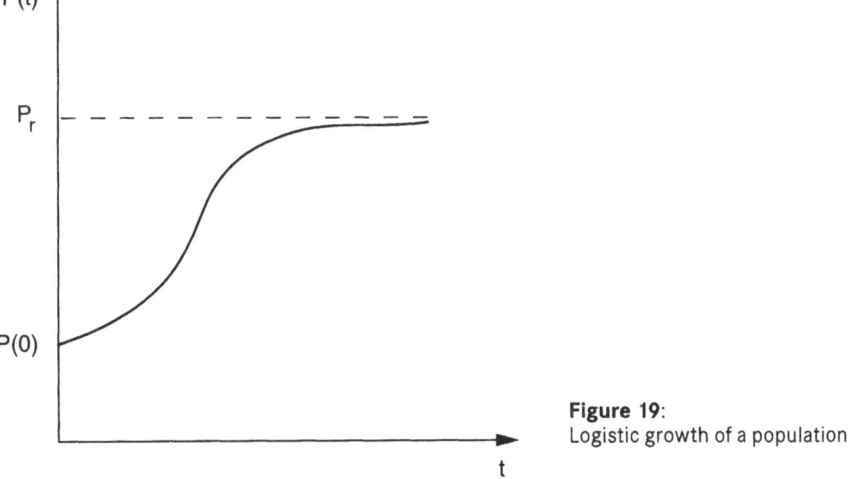

Figure 19:
Logistic growth of a population

and the (now nonlinear) feedback diagram takes the form of fig. 18. The population increases with time according to the curve (called "logistic") shown in fig. 19 which is characterized by an almost exponentially increasing first part and by a final part tending asymptotically to P_r.

Models of the type just considered form the basis of the analysis of many phenomena in ecology concerning the interaction of two populations between which
different kinds of relationships may take place (e.g., preying, parasitism and cooperation).

For studying these problems too, linear models could be employed by assuming
that the rate of change of each population is proportional not only to its level but
also to the level of the other population (through a coefficient that can be either
positive or negative). The relevant block diagram is represented in fig. 20, where
the two populations are denoted by P_1 and P_2. This diagram accounts for the variations of P_1 and P_2 by means of the feedback connection of two models of the type
shown in fig. 16 b). The response of the resulting (linear) second-order system contains, besides a constant term corresponding to the (constant) carrier capacity r,
pairs of exponentially increasing and/or decreasing modes, or an oscillatory mode
that may be damped, undamped or self-exciting (let us observe, in this regard, that
quantities P_i are intrinsically positive, whereas their variations may be positive or
negative). The shape of the response depends on the sign and values of the four
coefficients k_{ij}. Fig. 20 refers to the case in which P_1 is the predator population and
P_2 the prey population with carrier capacity r: persistent oscillations occur if $k_{11} =
k_{22}$ and $k_{12} k_{21} > k_{11} k_{22} = k_{11}^2$.

A more effective analysis, however, entails the resort to nonlinear models, like
those of Vito Volterra (1860–1940) and Alfred J. Lotka (1880–1949). The most
elementary model of this kind is shown in fig. 21, where the signs at the summing
points correspond to the case in which P_1 is the population of predators. Clearly,
an oscillatory behavior with P_1 and P_2 in quadrature is represented on the phase
plane by a centre in the linear case of fig. 20 and by a limit cycle in the nonlinear
case of fig. 21.

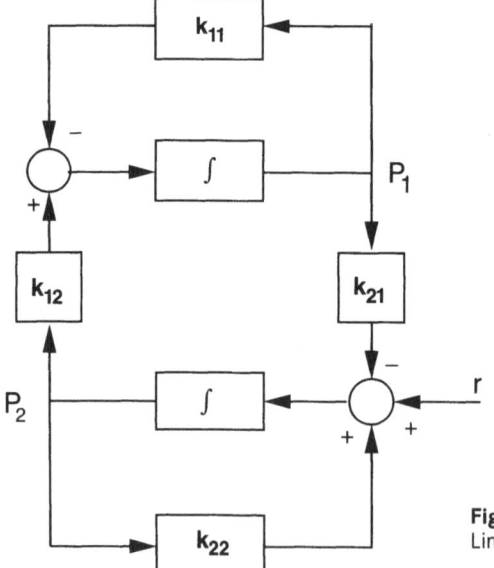

Figure 20:
Linear model of the coevolution of two populations

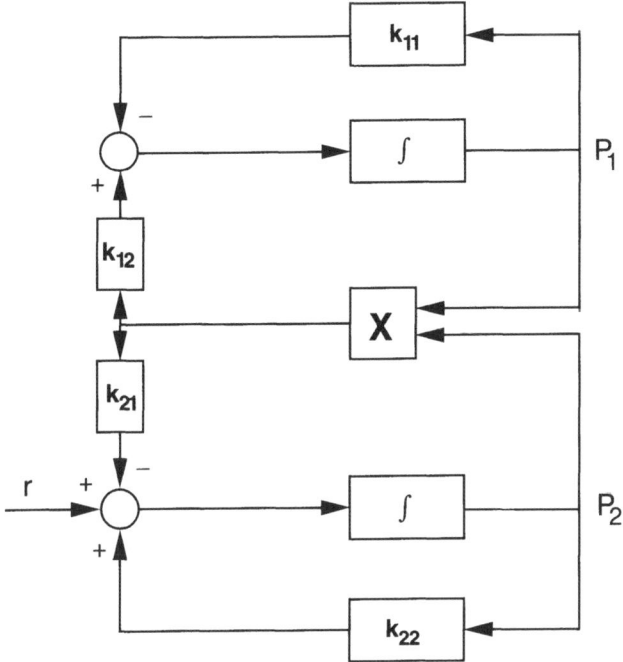

Figure 21:
Simple nonlinear model of the coevolution of two populations

Feedback Models in the Sciences of Man and Society

Let us first refer to a hard science like mathematical economics. A number of feedback models have been suggested to describe economic phenomena (for example Samuelson 1939, Hicks 1950, Phillips 1954, 1957). Here we focus on the simplest version of Alban William Phillips' (1914–1975) model, whose block diagram is represented in fig. 22. It shows how the aggregated demand Z affects production Y through a suitable functional relation D; in turn, Z is the sum of an exogenous term E (corresponding to governmental interventions), of the term C accounting for consumption and of the term I accounting for investments. Consumption is influenced by production through a coefficient K_C (propensity to consume) and functional relation D_C, whereas investments are influenced by the rate of change of Y through a coefficient K_I and functional relation D_I.

When $E = 0$, the two feedback channels (corresponding to the so-called multiplier and accelerator phenomena) may give rise to an oscillatory behavior and can be considered as possible causes of the economic cycle. On the basis of his model (incorporating intrinsic feeback loops) Phillips suggested an extrinsic feedback control system which supplies the time course of E capable of avoiding an oscillatory course of Y.

Feedback models of other interesting economic phenomena have been proposed by Jay W. Forrester (1961, 1969). They are based on the methodology called "system dynamics" (Forrester 1968), which exploits the feedback connection of elementary models like that in fig. 23: a *level variable L* is computed as the integral of the

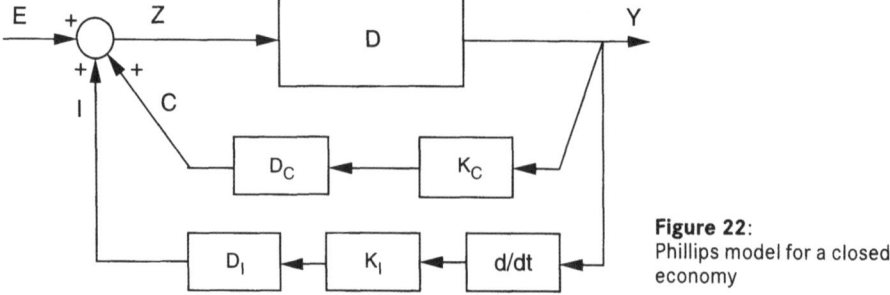

Figure 22:
Phillips model for a closed economy

difference of an inflow F_i and an outflow F_o (flow variables); these flows may be influenced (as flow F_i in fig. 23) by a level variable through a *decision function* D. The block diagram corresponding to the system dynamics diagram of fig. 23 is drawn in fig. 24 (Cobelli, Lepschy, Romanin Jacur, and Viaro 1986). Note that, in the specific case, F_o flows out from the reservoir (as fig. 23 points out), whereas its variation is independent of L (because it is due to an external action, e.g., that of a pump) and affects the level and, consequently, the inflow (as fig. 24 points out).

More interesting are the feedback models in the so-called soft sciences, in which a formal and quantitative approach is not usual. In this field we shall mention first the model of Frederick William Lanchester (1868-1946) concerning the evolution of military conflicts (Lanchester 1956). In its simplest version it is assumed that each army suffers losses proportional to the size of the opposing army; therefore, denoting by y_1 and y_2 the sizes of the two armies, their dynamics is described by the system of differential equations:

[6]
$$\begin{cases} \dfrac{dy_1}{dt} = -k_2 y_2 + r_1 \\ \dfrac{dy_2}{dt} = -k_1 y_1 + r_2 \end{cases}$$

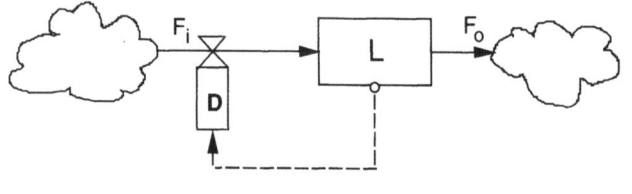

Figure 23:
Basic elements of the "system dynamics" schematics

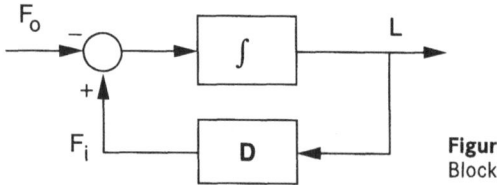

Figure 24:
Block diagram corresponding to the model of fig. 23

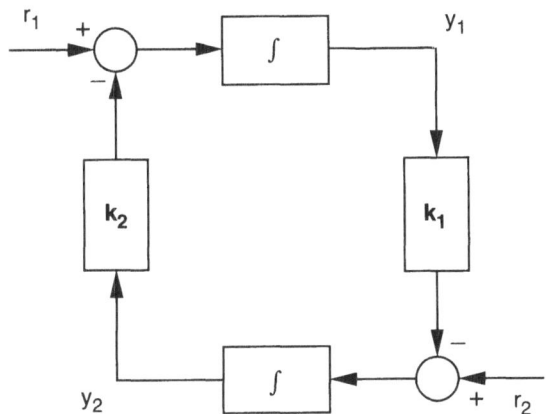

Figure 25:
Lanchester model for a military conflict

where coefficients k_1 and k_2 are positive numbers related to the effectiveness of the respective armaments and r_1, r_2 are inputs for replacing the losses. The related block diagram exhibits a feedback loop (fig. 25). The same block diagram could represent phenomena of cooperation or coevolution by attributing different signs to coefficients k_1 and k_2.

Another interesting model has been suggested by Lewis Fry Richardson (1881–1953) to represent the armament race in quantitative terms (Richardson 1960). By limiting attention to the two-power case and indicating their armament levels with y_1 and y_2, the simplest form of the model is the following:

[7]
$$\begin{cases} \dfrac{dy_1}{dt} = k_{12}y_2 - k_{11}y_1 + c_1 \\ \dfrac{dy_2}{dt} = k_{21}y_1 - k_{22}y_2 + c_2 \end{cases}$$

where coefficients k_{ij} are called *defense* coefficients, coefficients k_{ii}, *fatigue-and-expense* coefficients, and coefficients c_i, *attitude* coefficients; the last ones are positive in the case of an antagonistic attitude and negative in the case of a cooperative attitude ("positive" and "negative" obviously refer to the algebraic and not to the ethical meaning of the terms). In this case, the block diagram exhibits three interconnected feedback loops, as shown in fig. 26 which has the same structure as fig. 20.

Many examples of feedback models in the field of social sciences are presented and discussed in the essays collected by John H. Milsum (1968). In the studies of Magoroh Maruyama (1961, 1962) on the Danish national character and on the algebra of interpersonal relations, he considers attitudes which favour other attitudes in a direct or, more often, mediated way; therefore, loops are present which entail a mutual strengthening of these attitudes, thus forming the typical "character" of a more or less numerous group of people (in this particular case, an entire nation).

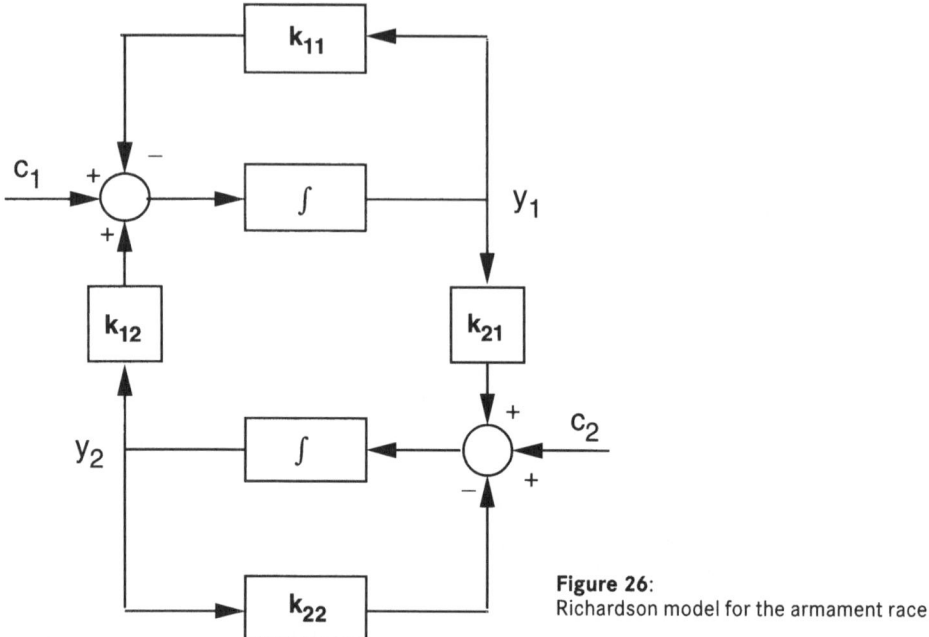

Figure 26:
Richardson model for the armament race

It seems reasonable to mention in this context the so-called world dynamics of Forrester (1971), even if the objectives pursued by the author have not been achieved. Forrester's approach is based on a rather simple, yet realistic, representation of the growth of world population, the exploitation of non-renewable resources, pollution, and industrial production, and on the analysis of their mutual relations. This model provides the structure of the relevant equations, but their solution requires the precise knowledge of their parameters. To this purpose, plausible values for them were first considered and, then, they were adjusted properly to account for the actual evolution of the above variables in the period 1900–1970 (for which reliable data were available).

The model identified in this way was used to forecast the course of the four relevant variables, as well as that of a conventional index for the quality of life, corresponding to different intervention policies (with particular reference to the exploitation of natural resources and demography). The resulting forecasts do not agree with the actual evolution after 1970. Nevertheless, Forrester's approach proves useful in practice, as it allows us to realize that, in some cases, apparently effective interventions do not produce the desired results, because the interactions among the variables involved, often related to the presence of feedback loops, give rise to "counterintuitive" behavior of the overall system.

Feedback Models and Philosophical Problems

It has often been noted that feedback models allow a mechanistic interpretation (where the effects depend only on efficient causes, to use a traditional term of

Aristotelian derivation) of phenomena exhibiting, at least at first sight, a finalistic (or goal-oriented) behavior (where the effects appear to be influenced by final causes). This typically occurs in the feedback models of many biological systems, in particular those regarding homeostasis and tropisms.

Feedback can, furthermore, offer suggestions on how to formalize problems that are the immediate object of a philosophical reflection. This will be illustrated with reference to the Hegelian dialectics of thesis, antithesis and synthesis. According to its vulgate version, dialectics consists in: posing an "abstract and limited" concept (thesis), suppressing this concept as something finite and going over to its opposite (antithesis), and finally arriving at the *synthesis* of the preceding determinations, which retains what in each one is valid and, at the same time, surpasses them both (the German term used by Georg Wilhelm Friedrich Hegel (1770–1831) is *aufheben*, which has been rendered as 'sublate'). Hegel actually writes about the intellectual, dialectic and speculative (or rational positive) stages but dialectics is not linked to the second stage only, but rather to the whole process and, owing to the identity of rational and real, it is the key not only of scientific progress but also of the "becoming" of reality. Accordingly, reality develops dialectically, so that triads of thesis, antithesis and synthesis are perceived everywhere.

This is not the place to discuss whether, or how far, this presentation corresponds to Hegel's thought. However things may be, this "popular" version can be presented in graphic terms as in fig. 27a), where the process that leads from thesis T to antithesis A is represented by block S_1 and the one leading from thesis and

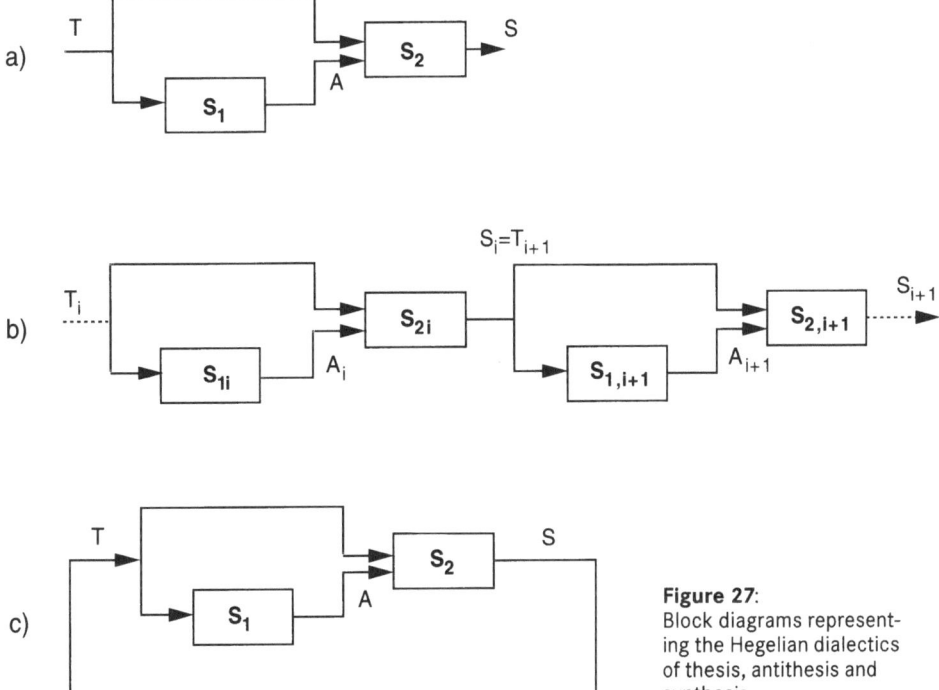

Figure 27:
Block diagrams representing the Hegelian dialectics of thesis, antithesis and synthesis

antithesis to synthesis S is represented by block S_2. Dialectic process, however, is something continuous: the synthesis arrived at in the first phase reproposes itself as a new thesis, which generates a new antithesis and combines with it, giving rise to another synthesis, and so on. The situation, outlined in fig. 27b) where the dashed line means that the chain should be prolonged indefinitely, is equal to the situation encountered in the already mentioned mechanical computer of W. Thomson and in the graphs studied by D. Scott that are equivalent to feedback loops (see section 6.3). The same procedure can also be adopted in the present case, thus arriving at the diagram of fig. 27c): synthesis S obtained according to fig. 27a) becomes the input of the same diagram as thesis T. In this way, feedback interprets the dialectic process satisfactorily. It accounts for that feature (not always easy to illustrate "in words") according to which the passage from thesis and antithesis to synthesis, the posing of the synthesis as a new thesis etc. are to be considered as moments not of a *temporal* but of a *logical* process. At the same time, the feedback model of fig. 27c) allows us to understand the compatibility of such point of view with the cases in which, in the historical development, one passes, through temporarily distinct and successive phases, from a position initially adopted to another position which exhibits contrasting and opposing characteristics, until one finally tends towards a more balanced, in a sense intermediate, position. In fact, such a behavior characterizes many feedback systems where the evolution of the output exhibits damped oscillations: a feature that is quite familiar to those who are concerned with feedback control systems.

Feedback may profitably be used to deal in a more satisfactory way with other problems of interest in philosophy. Let us consider, for instance, a classic dispute regarding the methodology of science. According to the naïve inductivism, induction leads from the observed facts $O.F.$ to the theory T, and deduction allows us to make forecasts F and derive explanations E: this situation may be represented with the open-loop diagram of fig. 28a).

It has been objected, however, that observations depend on a theory, propositions describing observations presuppose it, and experiments are guided by it. Therefore, one may reasonably resort to the feedback diagram of fig. 28b) in which two distinct inputs are present, i.e., the spontaneous observations $S.O.$ and the preliminary ideas $P.I.$ on the related phenomena (which are often unconscious). The observed facts $O.F.$ not only come from spontaneous observations but also from the experiments Ex performed according to the theory T; the last, in turn, is enriched, modified or, if necessary, replaced by a different theory suggested by the reflections R on the observed facts. The diagram of fig. 28b) seems to be particularly interesting since it accounts both for aspects that are more directly related to the methodology of science and for aspects that are connected with the psychology of discovery.

Of course, a very simple diagram may not fully explain the general features of the phenomena under investigation (in other words, it may be regarded as a simple explanatory model); a more detailed description requires more complex diagrams whose kernel, however, is the simple model we have considered.

These observations also apply to the following example concerning the hypotheses on the relationships between language and thought or between language

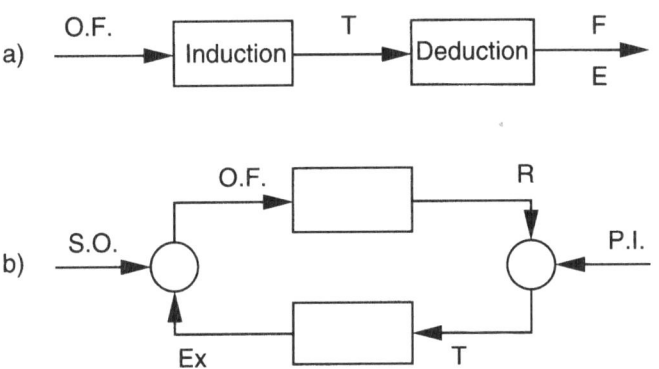

Figure 28: Block diagrams for representing: a) the position of the naïve inductivism, b) a position claiming mutual influence between induction and deduction

and culture. With a rough simplification, four hypotheses may be considered: (a) the so-called strong version of the Edward Sapir (1884–1939)-Benjamin Whorf (1897–1941) hypothesis (or, rather, a somewhat arbitrary extension of it) according to which language itself gives shape to ideas (different *Weltanschauungen* correspond to different languages); (b) the strong version of the Jean Piaget (1896–1980) hypothesis (cognitive hypothesis) according to which language is structured by thought and reflects it; (c) the genetic-evolutionary hypothesis of Lev Semonovich Vygotsky (1896–1934) for which thought and language develop simultaneously through continual mutual influences; and (d) the Franz Boas (1884–1942) hypothesis which claims the substantial independence and the absence of a tight and necessary relationship between thought and language.

If an eclectic approach is adopted, the feedback diagram of fig. 29 is obtained: here two independent inputs are present (hypothesis of Boas), corresponding, respectively, to the component L' of language irreducible to the influence of culture C and to the component T of thought irreducible to the influences of language. Language, however, contributes a component C'' to culture (hypothesis of Sapir-Whorf) through a feedback path in the same way as culture contributes a component L'' (hypothesis of Piaget) through a complementary feedback path. Finally, the feedback structure of the overall diagram turns out to be consistent with the ideas of Vygotsky.

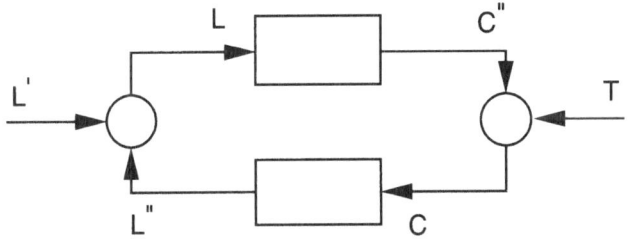

Figure 29: Feedback block diagram representing the relation between language and thought according to an eclectic approach

Feedback models can also be employed to represent the mutual influence between the meaning of a sentence and the meaning of its component parts. As is known, in the formal languages the semantic "principle of compositionality" holds. It may be stated as follows: "The meaning of a sentence is determined by the meaning of its words (and by its structure)". In natural languages, instead, most words have several meanings. Therefore, the "principle of semantic holism" holds for them. It may be stated as follows: "The meaning of a sentence determines the meaning of its words". In deciphering texts written in a partially unknown language, as well as in other analogous problems, both the above principles can be applied.

The initial idea about the meaning of some words influences the meaning attributed to the entire sentence and this new, or more complete, meaning enriches, or makes more precise, the meaning of its words. This process can be interpreted by means of the diagram of fig. 30: the circular blocks account for the comparison between the meaning previously assigned to a word and those meanings of the word that are compatible with the new meaning of the sentence and for the comparison between the meaning previously assigned to the entire sentence and the meanings now assigned to its words.

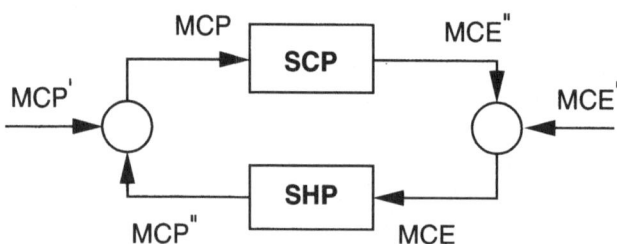

Figure 30: Feedback block diagram representing the mutual influences between the semantic compositionality principle SCP and the semantic holistic principle SHP: MCP': preliminary meaning attributed to each component part of a sentence; MCP": contribution to the meaning of each component part depending on the meaning of the whole sentence; MCP: resulting meaning of each component part; MCE': preliminary meaning attributed to the whole sentence; MCE": contribution to the meaning of the whole sentence depending on the meaning of its component parts; MCE: resulting meaning of the whole sentence

A final example refers to the evolutionary process that has led to the adaptation of human beings to changing environments through a number of nested control systems, as shown in fig. 31. This combines several diagrams suggested by Edoardo Boncinelli (2000). The reason why this model has been included in the present section is that each control loop concerns a different scientific area; at the lowest level physical and chemical aspects are considered, whereas at the highest level cognitive and social aspects are taken into account. The consideration of all the above-mentioned aspects in the same diagram is undoubtedly of interest from the epistemological point of view.

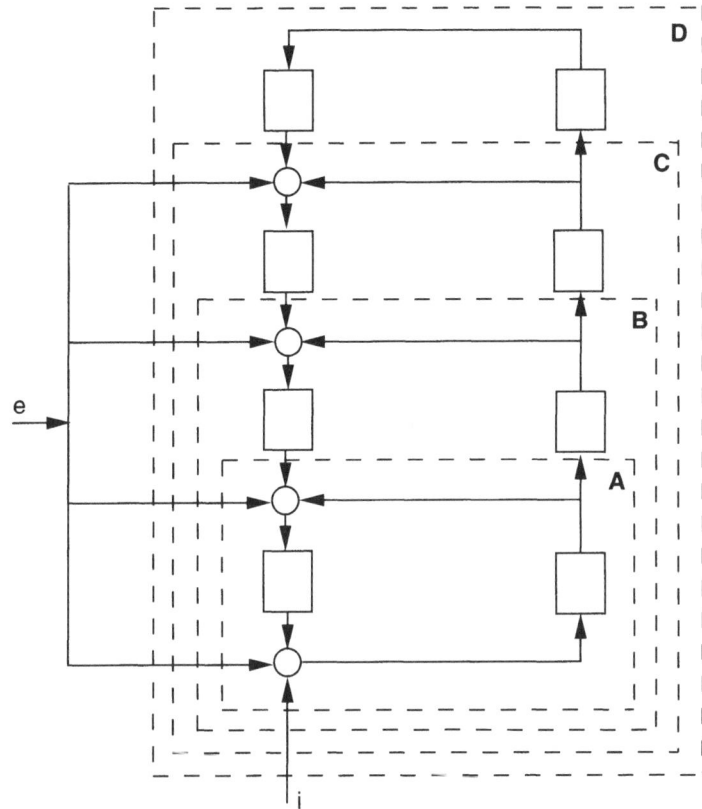

Figure 31: Nested feedback loops accounting for evolutionary adaptation; A: genetic control loop (adaptation to local perturbations during the growth); B: genomic control loop (adaptation to intergenerational changes); C: cognitive control loop (adaptation to changes occurring during the individual life); D: socio-cognitive control loop (adaptation to the socio-cooperative action); i: information supplied by a generation to the next (adaptation to transgenerational changes); e: environmental perturbations

6.5 Concluding Remarks

It will not escape the reader that the block diagrams of figs. 28b), 29, 30 have the same structure, which is substantially similar to that of fig. 25. In these cases, in fact, two *equally important* phenomena influence each other, i.e., the evolution of two distinct populations in fig. 25, the inductive and deductive processes in fig. 28b), the language-culture and culture-language relations in fig. 29 and the part-whole and whole-part interactions in understanding a sentence in fig. 30.

A different situation characterizes fig. 16b), where an *intrinsic feedback* is present. The same diagram arises in the artificial systems in which an *extrinsic feedback* channel is intentionally adopted to exert a control action or to modify the system performance. In these cases, the feedback channel plays an *auxiliary* role with respect to the controlled process.

154 Antonio Lepschy and Umberto Viaro

Of course, more complicated situations may occur. For instance, fig. 20 corresponds to the mutual influence of two phenomena each characterized by an intrinsic feedback, whereas fig. 22 exhibits two intrinsic feedback channels. Even more complicated are the arrangements of figs. 7 and 15.

The variety of situations considered in this paper clearly shows the usefulness of the feedback concept which can effectively be exploited in the technological field, as well as in mathematics and in the construction of simple explanatory models for a large class of phenomena belonging to both the "hard" and the "soft" sciences.

References

Boncinelli, E., 2000, *Le Forme della Vita. L'Evoluzione e l'Origine dell'Uomo*, Torino, Einaudi.

Cobelli, C., Lepschy, A., Romanin Jacur, G., and Viaro, U., 1986, "On the relationship between Forrester's schematics and compartmental systems", *IEEE Transactions Systems, Man, and Cybernetics*, 16: 723–726.

Forrester, J. W., 1961, *Industrial Dynamics*, Cambridge (Mass.), MIT Press.

Forrester, J. W., 1968, *Principles of Systems*, Cambridge (Mass.), MIT Press.

Forrester, J. W., 1969, *Urban Dynamics*, Cambridge (Mass.), MIT Press.

Forrester, J. W., 1971, *World Dynamics*, Cambridge (Mass.), MIT Press.

Hicks, J. R., 1950, *A Contribution to the Theory of the Trade Cycle*, London, Clarendon Press.

Lanchester, F. W., 1916, *Aircraft in Warfare: The Dawn of the Fourth Arm*, London, Constable.

Lanchester, F. W., 1956, "Mathematics in warfare", in: *The World of Mathematics* (Newman, J. R., ed.), vol. 4, New York, Simon & Schuster: 2138–2157.

Lepschy, A., Mian, G. A., and Viaro, U., 1992, "Feedback control in ancient water and mechanical clocks", *IEEE Transactions on Education*, 35: 3–10.

Maruyama, M., 1961, "The multilateral mutual causal relationships among the modes of communication, sociometric pattern and the intellectual orientation in the Danish culture", *Phylon*, 22: 41–58.

Maruyama, M., 1962, "Algebra of interpersonal interaction", *Methodos*, 13: 25–36, 49–50.

Maxwell, J. C., 1868, "On Governors", *Proceedings of the Royal Society*, 16: 270–283.

Mayr, O., 1970, *The Origins of Feedback Control*, Cambridge (Mass.), MIT Press.

Milsum, J. H. (ed.), 1968, *Positive Feedback. A General Systems Approach to Positive/ Negative Feedback and Mutual Causality*, Oxford-New York, Pergamon Press.

Phillips, A. W., 1954, "Stabilisation policy in a closed economy", *The Economic Journal*, 64: 290–323.

Phillips, A. W., 1957, "Stabilisation policy and time-forms of lagged responses", *The Economic Journal*, 67: 265–277.

Richardson, L. F., 1960, *Arms and Insecurity: A Mathematical Study of the Causes and Origins of War*, London, Stevens.

Routh, E. J., 1877, *A Treatise on the Stability of Motion*, London, Macmillan.

Samuelson, P., 1939, "Interaction between the multiplier analysis and the principle of accelerator", *Review of Economic Statistics*, 21: 75–79.

Scott, D., 1971, "The lattice of flow diagrams", in: *Symposium on Semantics of Algorithmic Languages* (Engeler, E., ed.), Berlin, Springer: 311–372.

Thomson, W., 1876a, "Mechanical integration of the linear differential equation of the second order with variable coefficients", *Proceedings of the Royal Society*, 24: 269–271.

Thomson, W., 1876b, "Mechanical integration of general differential equation of any order with variable coefficients", *Proceedings of the Royal Society*, 24: 271–275 (including an "Addendum", received February 10, 1876).

7 Adequacy of Mathematical Models in Control Theory, Physics, and Environmental Science

EVGENII F. MISHCHENKO, ALEXANDR S. MISHCHENKO AND MIKHAIL I. ZELIKIN

In contemporary Russian, the word "model" is often associated with a showing of fashionable clothes, that is, with something purely external having no more purpose than to decorate the real human essence. Here, we discuss the very opposite aspect of this notion. When speaking of mathematical models, we imply a speculative construction designed to express the real essence of a phenomenon and the causes of the processes in question. It must be clearly perceived that all natural phenomena are interrelated. The modest abilities of the human intellect do not allow taking all relations into account. But, luckily, some relations are strong, and others are vanishing. So we have a possibility to reveal the main acting forces, discarding all that are of secondary importance. This is the common basic paradigm of all natural sciences, if we exclude their purely descriptive aspects. This paradigm preordains the leading role of mathematics in the process of designing and exploring models.

We attempt a justification of this thesis. The question is not only in the unbreakable logic and rigor of mathematical reasoning. It would be natural to ask: Is there a real need for rigorous mathematical proofs, which mathematicians use even when the fact in question is self-evident? It is generally believed that we need proofs to be absolutely convinced of the correctness of statements. In a sense, this is true, and there are numerous examples of statements that seemed true but long defied proof until refuting counterexamples were found. Meanwhile, this is only a surface aspect of a sufficiently deep problem. Strict proof of a purely mathematical theorem is only the process of deduction from a set of axioms accepted (often implicitly) during a given period of mathematical development. But according to Gödel's theorem, in any given sufficiently rich axiomatic system, there exists a proposition that can be neither proved nor disproved within this system. That is why, in general, it is impossible to conclusively determine in which extension of the commonly accepted axioms the theorem is true. Hence, its absolute validity always remains problematic. Aristotle had an interesting point of view on this problem. He thought that proofs are needed to reveal the reasons for mathematical facts. Thus, in accordance with Aristotle's views, even purely abstract mathematics itself turns out to be designing models (in a certain higher sense). But these models are not constructed independently of each other. They are organized in an well-integrated Whole: a most sophisticated system of views or (figuratively speaking) a beautiful architectural building – the building of Mathematics. Namely the incorporation into that well-integrated Whole gives us intuitive trust in the validity of mathematical statements. Because mathematics is the most ancient and the most perfect among the deductive sciences, its experience in creating models can serve by right as a pattern for imitation.

When it comes to a concrete problem (in physics, biology, geography, etc.), we need the intuition of a scientist who has long explored the object in question, and it is necessary to know facts specific to this branch of science. Good models appear as a result of contacts between professional mathematicians and experts in the corresponding profession. For instance, the united efforts of the economist John Maynard Keynes (1883–1946) and the mathematician Frank P. Ramsey (1903–1930) allowed formulating and proving "the golden rule of accumulation" in mathematical economics. The collaboration of the mathematician John von Neumann (1903–1957) and the economist Oskar Morgenstern (1902–1977) provided the base for applying game theory to mathematical economics. An amazing example is the alliance of scientists who created quantum mechanics. They were physicists, but they constituted a highly heterogeneous collective: Arnold Sommerfeld (1868–1951) had considerable mathematical experience, while Werner Heisenberg (1901–1976) had a unique physical intuition. There are examples of indirect cooperation. For instance, the mathematicians Henri Poincaré (1854–1912), David Hilbert (1862–1943), and Hermann Minkowski (1864–1909) played a fundamental role in creating relativity theory hand in hand with the physicists Hendrik A. Lorentz (1853–1928) and Albert Einstein (1879–1955). And the first rigorous mathematical model of quantum mechanics was constructed by von Neumann. Unfortunately, this cooperation is often prevented by the personal ambitions of scientists who fail to properly value the contribution of their colleagues.

The authors of this paper were fortunate to learn and work under the direct guidance of a great mathematician of the 20th century, Lev Semenovich Pontryagin (1908–1988). The oldest of us, Evgenii Frolovich Mishchenko, together with Revaz Valerianovich Gamkrelidze (b. 1927) and Vladimir Grigor'evich Boltyanskii (b. 1925), played an outstanding role in the birth and establishment of the widely known Pontryagin mathematical school in control theory and differential equations. This school became famous among mathematicians and engineers immediately after the discovery of the maximum principle, now called Pontryagin's maximum principle. Mikhail Il'ich Zelikin was then a student whose scientific adviser was Pontryagin and Alexandr Sergeevich Mishchenko collaborated with Pontryagin in the 1980s. At that time, we participated in the powerful environmental movement that arose in the USSR against the infamous, huge, nature-destroying projects for turning northern and Siberian rivers to the south. Having been affirmed by the Central Committee of the Soviet Union Communist Party, these projects in actuality had the force of law. After these projects were rescinded, Pontryagin organized a laboratory at the Steklov Mathematical Institute to investigate mathematical environmental problems (Zelikin and A. S. Mishchenko have an important role in this laboratory). At the same time, a nongovernmental public organization was organized – the Association of Scientists "Ecology and World" – uniting scientists (among them outstanding members of the USSR Academy of Sciences) who were anxious about environmental problems; A. S. Mishchenko became the vice president of the Association and Zelikin was a member of the board. In the course of the active work of these organizations, a large number of environmentally dangerous national-economic projects were scrutinized, and scientific assessments of their justification and probable impact were reported.

Our aim in this chapter is to share our experience in the mathematical modeling of technological and environmental processes and, based on this experience, to discuss the philosophical and epistemological aspects of modeling problems.

7.1 Mathematical Models of Technological Processes

We first want to describe Pontryagin's excellent achievements in the mathematical modeling of engineering and technological processes. For the most part, they relate to four areas: the concept of structurally stable dynamic systems, the theory of relaxation oscillations, Pontryagin's maximum principle, and the theory of differential games (in this paper, we omit Pontryagin's widely known results in algebra and topology).

Pontryagin investigated structurally stable dynamic systems in collaboration with the physicist Aleksandr A. Andronov (1901–1952). They were stimulated by the technological problem of the oscillation stability of railroad cars. At high train speeds, its cars begin to sway, and that swaying may result in a wreck. When a model is created, the functions describing the system in question are never known exactly. We need to understand in what cases the qualitative behavior of the system does not change with small variations of these functions. A system of differential equations is said to be structurally stable if for any sufficiently small change of the vector field defining this system, the phase portrait remains topologically invariant. Andronov and Pontryagin proved that any structurally stable two-dimensional system of differential equations has only nondegenerate singular points, nondegenerate limit cycles, and separatrices in the general position. After this pioneer work, many mathematicians tried to extend this result to the multi-dimensional case. Only after a long time, it was discovered that the situation is considerably more complicated in the multidimensional case. Dmitrii Victorovich Anosov (b. 1936), a student of Pontryagin, described a class of systems having a type of geodesic flow on a manifold of negative curvature that does not satisfy the above mentioned conditions. And the famous example of Stephen Smale (b. 1930), the so-called horseshoe of Smale, supplies mathematicians with a visual mathematical model of hyperbolic behavior. This model enriches the intuition of mathematicians and provides a base for far-reaching mathematical theories having considerable practical import. This shows that a specific exactly explored example can serve as an excellent model.

The theory of relaxation oscillations is connected with investigating the asymptotic behavior of systems of differential equations with a small parameter in the highest derivatives when this parameter tends to zero. Problems in radio engineering served as a stimulus for constructing this theory (see chapter 3). In the deduction of differential equations describing the operation of an electronic device, small parameters that are not specified but are nevertheless incorporated in the design (such as additional capacitance, inductance, etc.) must be neglected.

Will the differential equations obtained adequately describe the operation of the device? A relatively simple case is when the small parameter occurs in the right-hand side of the differential equations. In this case, we have Poincaré's theorem asserting that the solution is a holomorphic function of the parameter, and we can take the Taylor series describing the asymptotic behavior of the solution. The case of a singular perturbation when the highest derivatives are multiplied by a small parameter is much more difficult. In this case, we must compare differential equations acting in different spaces. We consider the larger of the two spaces and its submanifold obtained by setting the small parameter to zero. Balthazar L. van der Pol (1889–1959) discovered that the motion of the phase point in that submanifold, induced by the dynamics in the ambient space, can be discontinuous. He found a second-order differential equation with a discontinuous limit cycle. A similar situation was found in the equations for a vacuum tube oscillator. Pontryagin, together with E. Г. Mishchenko, found exact conditions for the existence of a multidimensional discontinuous limit cycle and calculated the asymptotic behavior of solutions of the differential equations in the neighborhood of this limit cycle. The corresponding series in a small parameter is radically different from a Taylor series. They not only proved that the limit behavior of the system is defined by a discontinuous limit cycle but also estimated the exactness of the approximation of the system's solution by its limit. E. F. Mishchenko, together with his students Nicolai Christovich Rozov and Andrei Yurievich Kolesov, later computed the full asymptotic series in a small parameter in the vicinity of the limit cycle.

One of Pontryagin's most famous achievements is Pontryagin's maximum principle. The history of its creation is instructive. To investigate a technological problem and design an adequate mathematical model of it, Pontryagin periodically listened to engineers and experts in remote control theory in his seminar. Aleksandr A. Feldbaum, Lev I. Rozonoer, Aleksandr Ya. Lerner, and many others spoke in this seminar.

Representatives of the applied sciences often have misleading views of using mathematics. They consider mathematics a collection of ready-made recipes: theories, mathematical facts, definitions, theorems, etc. Mathematicians, in their opinion, are librarians who know where to find the applicable mathematical result. But, in fact, designing a serious mathematical model is always a creative task requiring new mathematical ideas, new theorems, and, not rarely, entirely new mathematical concepts that require additional investigation. The role of professional mathematicians in this creative process is invaluable. Further, when scientists lacking an adequate mathematical qualification try to ask mathematical questions, they even seem unable to formulate it in the proper exact form. Finally, formulating a mathematical problem itself implies a simultaneous subconcious search for the mathematical tools appropriate to its solution. All these factors clarify why the alliance of mathematicians and representatives of other natural sciences seems so fruitful.

The realization and treatment of the concepts expressed (sometimes implicitly) by the speakers in the seminar was the groundwork for building the mathematical theory of optimal processes. Most important was the insight into the final result to be proved. Pontryagin had such perspicacity to the highest degree.

After the maximum principle was proved, there was a powerful resonance in scientific circles. Many applications to finding the optimal behavior of all sorts of objects appeared in very different branches of knowledge. To use the maximum principle, one need only formulate the problem in question as a controlled system of differential equations. The maximum principle thus becomes a model for constructing models in various natural sciences. The opposite reaction was a wave of criticism aimed at depreciating the significance of the maximum principle and preventing its overestimation in scientific circles. Some mathematicians assert that the maximum principle is only a modification of known classical constructions in the calculus of variations: the Lagrange multiplier rule and the Weierstrass optimality condition. But in so doing they consciously (maybe semiconsciously) forget that these classical constructions are applicable only if the control takes values in an open set (in the topological meaning of the word). And the power of the maximum principle is that it in fact works independently for any sets of control. On the other hand, representatives of the engineering sciences object to the maximum principle because the concept of instantaneous switching seems unrealistic to them from the practical standpoint. To overcome this objection, a number of theorems were proved showing that system solutions in which the switching process takes some small fixed time interval (other variants: is realized at a random instant in the vicinity of the switching point or is realized with a delay) asymptotically approach the system solutions with instantaneous switching as the switching interval tends to zero.

Almost half a century has passed since the maximum principle was established. During this sufficiently long period, Pontryagin's maximum principle has been used with great success, demonstrating its theoretical and applied significance.

In his last years, Pontryagin was involved in the theory of differential games, which he helped to create. The theory of differential games describes mathematical models of conflict situations in the case when the strategies are processes continuous in time. One of Pontryagin's brilliant achievements in this area of mathematics was constructing the method of Pontryagin's alternated integral. It is a beautiful abstract construction using the operation of Minkowski's geometric difference. With this construction, Pontryagin explicitly described the domains of initial points from which one player can inevitably overtake the other.

The main theorem in differential game theory – the saddle point theorem – was proved by another outstanding specialist in mathematical modeling, Nicolai Nicolaevich Krasovskii (b. 1924). To prove it, he was forced to introduce a far-reaching extension of the notion of a solution of a controlled system of differential equations with a discontinuous right-hand side. At the same time, such extended solutions are quite natural from the application standpoint because they take into account that real control influences in the presence of an opponent are in fact discrete.

Control with an Accumulation of Switchings

At the first International Federation of Automatic Control Congress in Moscow in 1961, at the same congress where Pontryagin spoke on the maximum principle, there was a report by the British mathematician and engineer Anthony Thomas Fuller. The theme of the report was a simple concrete example. Suppose we wish to minimize the mean-square deviation of a particle from the origin when this particle moves along a straight line obeying Newton law, subject to a force that does not exceed one in absolute value and can be taken arbitrarily. Fuller was interested in the question of which optimality criterion leads to the most efficient stabilization (from the engineering standpoint) of such a controlled body. As was known, if the time of transition to the origin is taken as the criterion, then there is at most one switching. One must accelerate the particle towards the origin and then set the brake with full intensity. The instant of switching from accelerating to braking must be chosen such that the particle stops exactly at the origin. Fuller found that if the mean-square deviation from the origin is taken as the optimality criterion, then this strategy is nonoptimal. It is necessary to switch a little later. For example, we consider an initial position $x_0 > 0$, $\dot{x}_0 = 0$. The optimal strategy of Fuller's problem has infinitely many switches and consists of an infinite number of cycles. On the first cycle, we begin with the force directed towards the origin and then switch to the opposite direction of force. The switching instant results in the particle stopping at a point $-qx_0$ past the origin. Because the particle goes past the origin, we call this "overregulation". The cycle is repeated for the newly generated initial position, and the particle stops successively at the points $-qx_0$, $-q^2x_0$, $-q^3x_0$, ... These points constitute an alternating convergent geometric progression. We thus have an infinite number of successive overregulations. Because the cycle durations also form a convergent geometric progression, the whole process takes a finite time. Finally, the particle stops at the origin. And this is the unique and exact solution of Fuller's problem. Such control behavior is called "chattering".

When published in 1961, Fuller's example provoked curiosity, but it was considered only "interesting" and was soon forgotten. But it was later shown (in works by Zelikin and his student Vladimir Fedorovich Borisov) that chattering is typical for nonlinear multidimensional controlled systems, which appear in control problems of mechanical systems. The exact conditions when the whole space is fibered by two-dimensional integral manifolds were found, and the picture of phase trajectories in each fiber is equivalent to the phase portrait of Fuller's problem. This sort of optimal synthesis appears in many optimality problems in robotics, in problems of controlling the rotation of a solid body, in problems of space navigation, electronics, mathematical economy, and others. All these problems naturally correspond (as is true for any mathematical model) to the idealized situation, but the exact chattering solution was found namely for this ideal statement. This solution supplies the investigator with an orientation and gives him the possibility of preliminary estimates.

Much like Pontryagin's maximum principle, Zelikin's works on chattering in mechanical systems provoked criticism: How can an infinite number of switchings be realized when controlling real mechanical objects? In response, Zelikin and Ludmila Filippovna Zelikina proved an approximation theorem with the following meaning. If N switchings are made instead of an infinite number in a problem with optimal chattering behavior, then the deviation of the functional from its optimal value is estimated by $Ae^{-\alpha N}$, where A and α are positive constants dependent only on the problem in question. They applied this theorem to the problem of finding exact constants in inequalities that estimate the derivatives of a function in terms of the norm of this function and the norms of its higher-order derivatives. These inequalities are a powerful mathematical apparatus for proving important results. Knowledge of these constants (called Kolmogorov constants) is necessary for mathematical purposes. It turns out that the functions for which these constants achieve their exact values are solutions of extremal problems with chattering. The above mentioned theorem allows calculating Kolmogorov constants with an arbitrary exactness. The exponential estimate in the considered case, for example, yields two exact digits after the decimal point for two switchings, three such digits for three switchings, and so on.

The chattering theorems cannot be deduced by using only the maximum principle. Its proofs demand nontrivial mathematical techniques. At the points where switchings accumulate, the system of differential equations for Pontryagin's maximum principle has a sufficiently complicated discontinuity; the system solutions do not have any arbitrarily small interval of continuity adjacent to such a point. The proofs use the blowing-up procedure of the Poincaré mapping of the switching surface into itself and showed the existence of hyperbolic singular points of the system obtained.

At first sight, the very existence of an infinite number of switchings is unrelated to real natural processes, being only an artefact of the modeling. Nevertheless, everybody has certainly observed things falling on the floor and bouncing more and more often until finally stopping.

We consider a simple mathematical model. Let two parallel walls move toward each other with a constant velocity, and let a perfect elastic ball bounce back and forth between these walls. We suppose that the velocity of the ball changes sign instantaneously at each bounce while remaining the same in absolute value. Then this ball bounces an infinite number of times up to the instant when it is compressed between the colliding walls. This situation is quite familiar to anyone who has tried to stop a Ping-Pong ball by pressing it against the table with a paddle. A real ball undergoes finite (although very many) bounces because the strike occurs in a nonzero time. The shorter the strike duration is, the greater the number of bounces. In the theoretical case of zero-duration strikes, the number of bounces becomes infinite in a finite time interval.

Ironically, almost the same situation was modeled by Enrico Fermi (1901–1954), John Pasta, and Stanislav Ulam (1909–1984) in the mid-1950s in Los Alamos using the first large (for that time) computer MANIAC. They wished to use MANIAC for peaceful purposes in the intervening periods between calculations related to creating the H-bomb. Incidentally, in another series of calculations, they in fact found the soliton-like behavior of a chain of oscillators experimentally but could not explain it theoretically .

Fermi, Pasta, and Ulam decided to test Fermi's idea for explaining the existence of cosmic particles with a huge kinetic energy experimentally. Fermi suggested that energy is accumulated as a result of a large number of collisions of the particle with galaxies. A collision with a galaxy moving in approximately the same direction as the particle has substantially less probability than with one moving in the opposite direction. The second type of collision must increase the energy of the particle after reflection. A sufficient number of such collisions might produce particles with a huge energy, as if galaxies played Ping-Pong using a cosmic particle as a ball. Fermi, Pasta, and Ulam performed a set of experiments modeling the movement of the particle between two walls moving periodically under the condition of perfect elastic reflection. Fermi's conjecture was not supported by these experiments. In this context, we note the works of Leonid Moiseevich Pustyl'nikov, who argued that a relativistic reflection law must lead to energy growth. This story shows the great significance of theoretical models, whose absence or weakness blinds the investigator, depriving him of the fruits of his labor despite brilliant plans and excellent experiments.

7.2 Mathematical Modeling of Environmental Processes: Fluctuations in the Level of the Caspian Sea

We provide some background of the important social movement in the USSR against diverting the northern and Siberian rivers to the south. The idea of diverting the rivers began to gather momentum in the 1930s. It was a rather primitive concept. There is plenty of sun and free labor but a lack of water in the South, and there is plenty of water but a lack of sun and free labor in the North. Being unable to change the course of the Sun, let's turn the northern rivers to run in the opposite direction! The idea remained latent for a long time. At a special session of the USSR Academy of Sciences called "The socialist reconstruction and development of the Volga–Caspian basin" in 1933, Gleb Maksimikianovich Krzhizhanovskii (1872–1959) formulated principles for the territorial redistribution of water resources in the European part of the USSR. The end of the 1930s was the time of constructing big storage lakes on the Upper Volga. The war delayed these works. The 1950s was the peak period of implementing the Volga flow regulation program. At the May (1966) Plenum, the Central Committee of the Soviet Union Communist Party adopted a wide development program for land reclamation that gave the status of law to the idea of turning the rivers. On 24 July 1970, the Central Committee of the Soviet Union Communist Party and the Council of Ministers of the USSR passed Resolution No. 612 "On the perspectives of developing land reclamation, regulation, and redistribution of river flows in 1971–1985". This resolution declared a primary task of accomplishing the diversion of 25 cubic kilometers of water by 1985. Based on this resolution, the Alekseevskii Institute Soyuzgiprovodkhoz of the Ministry of Water Works of the USSR obtained a work order for elaborating the technological-economic basis for diverting part of the flow of the northern and Siberian rivers to the south. In 1976, Soyuzgiprovodkhoz was appointed General Planner of the diversion projects. These works were included in the "Main directions of developing the national economy of the USSR in 1976–1980" and were affirmed by resolution of the 25th Party Congress. In 1978, the Institute Soyuzgiprovodkhoz was renamed the Chief Institute on Diverting and Redistributing Waters of the Northern and Siberian Rivers.

The standard payback period for projects was established as 8.5 years. Any project with a payback period greater than 11 years was considered absolutely unprofitable. But the payback periods for the river diversion projects could not be compressed into such a narrow frame despite all stretches hypothesizing fantastically rich harvests on the irrigated areas. Then a ruse was conceived. If the level of the Caspian Sea will fall considerably (in the opinion of the diversion adherents, this was inevitable because of diverting water for irrigation), then all the Caspian sturgeon stock will perish, and the country will lose the profits from the export of sturgeon and black caviar. This profit was sufficient to draw the project payback period to the needed 11 years. Thus, the economic justification of the project strongly depended on the forecasted falling of the Caspian Sea level. The Caspian Sea is a mysterious object, and it deserves a special description.

The Caspian Sea is the largest body of water not joined to the world ocean, and its level varies significantly depending on climatic conditions. Geologic data testifies that there were phases in the past when the sea reached modern Volgograd. On the other hand, there is evidence of human activity rather deep under the water indicating that there were phases when the sea stood almost 12 meters lower than now. The research on the Caspian sea, as well as many things in Russia, was initiated by Peter the Great, who, in particular, established the first water-measuring post for regular observation of the level of the Caspian Sea at Baku. Incidentally, Peter, the first Russian elected to the French Academy of Sciences, was elected for his study of the Caspian Sea. Observation of the level was stopped and reactivated time and again for various reasons. A rather reliable, uninterrupted series of observations began in the 1860s. Since then, there was a long period when the sea oscillated near the level –26 m B.S., i.e., 26 m below the level of the world ocean as measured at a water-measuring post on the Baltic Sea (the abbreviation B.S. denotes the Baltic System). Eight dry years from 1933 to 1940 caused a fall of the Caspian Sea level to –27.5 m B.S., and the level then continued to gradually decrease (especially during the period of constructing hydropower stations on the Volga). In 1978, the level reached –29 m B.S. and then began to increase. The rising phase continues to the present.

The sudden, drastic change of the Caspian Sea level during 1933–1940 significantly damaged the economy of the Caspian region because of the necessity to reorganize municipal and port services, to rebuild coastal and marine oil extraction facilities, and so on. It generated a need for long-range predictions of the Caspian Sea level.

The bibliography devoted to studies and forecasts of the Caspian Sea level numbers hundreds of items and is characterized by an extreme diversity of approaches and views. The Caspian Sea is a very large water-collecting object. The behavior of the Caspian Sea level integrates a great variety of factors. The influence of local factors is insignificant, but all essential global climatic, hydrologic, and hydrogeologic changes are reflected in its behavior. Wide experience in studying the Caspian sea is accumulated at present. It is both solid experimental material and a wide range of theoretical research.

The Caspian Sea problem is significant because its level relates to the organization of the coastal and port economy, to problems of building roads and hotels, to oil extraction, and so on. Unpredictable changes in the level lead to multimillion expenditures. Ichthyologists assert that a level below –28.5 m will affect adversely the sturgeon stock (which, however, is now in an extremely poor state because of an unsatisfactory environmental situation and poaching and not because of changes in the sea level). On the other hand, the level rising higher than –27 m will lead to flooding newly built up and exploited coasts. Therefore, reliable long-range forecasts of the Caspian Sea level have great national economic significance. For namely these reasons, problems of predicting the Caspian Sea level takes an important place in hydrology.

We attempt to describe the most essential and promising approaches to the problem of predicting the Caspian Sea level popularly, but briefly.

Connections with Solar Activity

Approaches connected with studying Earth–Sun ties are based on establishing correlations between solar activity and the Caspian Sea level or between solar activity and the average discharge of the Volga. Wolf numbers, which are defined as the average annual number of sunspots (with due regard to their clustering and surface area), are usually used to characterize the solar activity. As is known, a clear-cut 11-year cycle can be observed in the dynamics of Wolf numbers. In addition, there exists a long-term cycle, but we do not have the centuries of observations of the level necessary for using it. An 11-year moving average, which smooths the 11-year oscillation component of the Wolf numbers, is more often used. A direct correlation between solar activity and the Caspian Sea level has naturally been insignificant because the Caspian Sea, as already mentioned, integrates many diverse factors in its behavior. The connection between solar activity and the average discharge of rivers is a little more reasonable hypothesis, although here again it would be too optimistic to hope for a linear correlation.

The search for correlations between solar activity and climatic processes provokes scepticism, if not irony, from modern climatologists. The most serious objection is the rather low energy of such an influence (in comparison with global atmospheric processes) and also lack of any known physical mechanism for explaining this influence. But there are counterobjections: First, the study of sufficiently complex processes (in biological systems for example) shows that low-energy but high-information influences are often much more significant than high-energy ones. Second, sunspots arise not by themselves but as a result of some unknown physical processes whose energy can be much more than the radiation changes due to the spots themselves. The appearance of spots only informs us of the high-energy processes that cause these spots.

The lack of a direct physical mechanism explaining the influence of solar activity on climate is not surprising. The matter here is the sufficiently complicated dynamics, which may be described by differential equations (and perhaps by some other even more complicated mathematical apparatus), and it is difficult enough, if not impossible, to find descriptive mechanical arguments for such dynamics. However, the insignificant correlation between solar activity and hydrologic processes does not mean that solar activity does not influence these processes. On the contrary, there are serious foundations for thinking that these two processes are interlinked. As a confirmation, we mention the prognosis of the Caspian Sea level made by Boris Semënovich Eigenson (1906–1962) in 1957. This prognosis, made using only connections with Wolf numbers, was confirmed in general. Namely, this prognosis predicted that 1980–2000 will be years of a high Caspian Sea level, which will reach marks of the 1930s by the end of the indicated phase. And this happened. It is therefore necessary to admit that the prognosis taking only Wolf numbers into account reflects some substantial features of the phenomenon in question. We note that in the prognosis, Eigenson supposed that the solar activity period is equal to 11 years. However, it was rather recently established that the solar activity is characterized not only by Wolf numbers but also by the direction of the magnetic field vector. In this connection, it is necessary to consider a solar activity period equal to not 11 but 22 years because the magnetic field of the spots reverses its direction each 11 years. Taking this circumstance into account can essentially improve the outcomes of predictions using this method.

Recently, Andrei Sergeevich Monin turned to the same ideas. He considered a simplified dynamics of the three-body system consisting of the Sun, Jupiter, and Saturn, neglecting the remaining planets, which have essentially smaller masses. The simplification is the assumption that all three bodies move on not elliptical but circular orbits around the common center of gravity. In particular, the Sun then orbits a point approximately 1.5 solar radii from its center. Jupiter's orbital period is close to 11 years. The tidal wave inside the Sun explains the 11-year cycle of Wolf numbers (maybe this is the high-energy physical process that we mentioned above). The recurrence period for the configuration of all three bodies is approximately 120 years, and this explains the long-term cycle of Wolf numbers. Monin searched (with some success) for known climatic cycles with the indicated periods.

Connections with Other Geo-cosmic Relations

A direct logical continuation of the "solar approach" were attempts to take other geo-cosmic relations into account. Such an approach was advocated and brought to life, in particular, by I. V. Maximov and his students. This approach involves a wide spectrum of ideas. One of the basic ideas is that the climatic oscillations of the European part of Russia are intimately interlinked with changes in the characteristics of the Gulfstream current. The natural question is what causes the changes in the characteristics of ocean currents. Among the reasons of a planetary character, it is first necessary to mention the long-period component of ocean tides (the period of the movement of the nodal line of the lunar orbit is approximately 18.6 years) and the variation of centrifugal forces under the influence of the nutation of the Earth's rotation axis (the period of modifications of the potential of the relevant deformation force, which results from combining a 14-month and an annual wave, is approximately 6–7 years). We note that a 19-year periodic component in the oscillations of the Gulfstream velocity and temperature is clearly traced. In Maximov's school, all these factors were compared with changes in river discharges. However, this approach has been developed only at a qualitative level up to now. The really interesting, but mathematically difficult, problem is to calculate oscillations of the form of the world ocean under the action of these forces and also to calculate the relevant fluctuations of the basic currents (Gulfstream, Kurosio, etc.) that essentially influence the climate.

Connections with Atmospheric Circulation

Modifications of the activity of the world ocean influence the climate through the atmosphere. There are many prognoses based on planetary atmospheric processes. The most substantial contribution here belongs to Boris L'vovich Dzerdzeevskii, G. Ya. Vangenheim, Aleksandr Aleksandrovich Girs, et al., who developed a theory of types of atmospheric circulation in the northern hemisphere and related it to changes in river discharges. Using this data, they forecasted a change in the Caspian Sea level. In particular, these scientists showed that a dominant circulation of the east type leads to aridification (i.e., to dry seasons), while a meridional or, especially, western type of circulation leads to humidification.

This approach was most fully expressed in the work of Klavdia Ivanovna Smirnova, who, in the course of many years at the Hydrometeorological Center of the USSR, successfully developed operating and long-term prognoses of the Caspian Sea level. We managed to find her. She had been dismissed on pension, to all appearances, not without the help of employees of the Ministry of Water Works, who needed absolutely different prognoses. When we asked her about forecasting methodology, she replied with embarassment that she did not use serious science. Her method was more similar to intuitive guessing based on knowledge and rich experience. She said that she had been under constant stress because economic decisions depended heavily on her prognoses. In the event of failure (in Stalin's time), she could quite possibly end up, as is said, "in places not so remote".

She had made a superlong-term prognosis for 19 years ahead. In connection with the 19-year oscillation period of the nodal line of the lunar orbit, a 19-year moving average was used to smooth such oscillations. The method was based on establishing a multiple correlation of changes of river discharges into the Caspian Sea with an index of atmospheric circulation (that of Girs, of Vangenheim, of Belinskii, and of others) and with atmospheric pressure fields. To estimate the pressure field contributions, the values were taken at the most typical points (in the sense of cyclonic activity) such as the Azores maximum, the Icelandic minimum, and so on. Taking these factors into account, she constructed a 19-year prognosis. Smirnova's prognosis, made at the beginning of the 1970s, had a good accuracy. The break from falling to rising of the Caspian Sea level in 1977 was predicted in this prognosis. The real and prognosticated levels differed in magnitude by no more than 30 cm. The high accuracy of her prognosis shows that methods based on atmospheric circulation and atmospheric pressure fields are promising.

Connections with Tectonic Activity

We now discuss approaches connected with the possible influence of the lithosphere on the Caspian Sea level. The underground component of drain into the Caspian Sea is usually considered stationary and unimportant, about 5 cubic kilometers annually (for comparison, the average annual discharge of the Volga is 250 cubic kilometers). The underground drain, in contrast to the river discharge for example, is not measurable by inspection. There are two standard explanations for fluctuation in the sea level. The first, conventional, is climatic: changes in the river discharges, in evaporation, and in precipitation. The second is tectonic: changes in the form of the Caspian Sea bed under tectonic influences. The second explanation is not established experimentally and has few supporters. The beautiful tectonic idea offered by Veniamin Petrovich Myasnikov is completely different. A huge amount of water is contained in pores of the ground and rocks of the Caspian Sea bed, several times the total amount of water in the Caspian Sea itself. There is some evidence to suppose that compression-rarefaction waves sometimes propagate in the Earth's crust with a very long period, on the order of decades. If so, a compression wave, reaching the Caspian sea, will squeeze out water from the ground pores, as from a sponge, into the sea, essentially increasing the input side of the water balance. A rarefaction wave, in contrast, will result in sucking water from the sea and accordingly in lowering its level. If it were possible to calculate and to predict such compression-rarefaction waves, it would give an effective method for predicting the tendencies of changes in level. It is clear that using only the theory of elasticity is inadequate here because the rate of propagation of elastic waves in the Earth's crust is very high. It is therefore necessary to take the influence of magma having, apparently, major viscosity into account and to try to find something like a seiche in the magma, which could initiate long-period crustal oscillations.

Periodic Components

In the previous sections, we   already discussed periodic processes apparently influencing fluctuations of the Caspian Sea level, such as the 11-year and long-term solar activity cycle and the 19-year cycle of rotation of the nodal line of the lunar orbit. There was a series of works where the authors attempted to isolate relevant periodic components in fluctuations of the Caspian Sea level and in elements of its water balance. Lev Semenovich Berg (1876–1950) and Arsenii Vladimirovich Shnitnikov took a priori given periods as a starting point, while B. A. Shlyamin examined the spectral expansion of level fluctuations and compared the frequencies found with the periods of external influences. After such analysis, the prediction obtained agreed with the real behavior fairly well. It seems rather probable that the level fluctuation (maybe, it would be better to consider the process of volume increments taking changes in the magnitude of the water surface into account) really represents a certain almost-periodic function. However, interpretation of its base frequencies should be made not a posteriori but in the course of modeling. That would unite the technique of Shlyamin and the approaches of Berg and Shnitnikov. Moreover, it would be necessary to evaluate beforehand the comparative potency of these and other actions.

Nonlinear Effects

When Zelikin and Zelikina drew and studied the monthly graphs of fluctuations of the Caspian Sea level, they noticed that the sea frequently stays at some selected or, as they used to call it, favorite levels. It was assumed that marks relevant to those sea levels at which the surface of shallow places of the sea attains a local maximum should be stable because a rise of the level a little higher than this mark will yield a sharp increase in the evaporating volume and consequently in the volume of evaporated water; conversely, a lowering of the level will yield a sharp decrease in the volume of evaporated water. The hypothesis was advanced that the equations describing the change of the level should be nonlinear and that the nonlinearity entails a composite asymptotic behavior: a random walk between favorite levels. This hypothesis has received partial confirmation in the work of Alexandr Sergeevich Mishchenko. He proved that the solution of the differential stochastic equation with cubic nonlinearity describing the water balance of the Caspian Sea has no more than three stationary solutions, of which two are stable and one is unstable. Any solution of the differential equation tends to one of the stable solutions. We recently clarified that if the degree of the polynomial is greater than three, then the differential stochastic equation can have as many stable stationary solutions as desired. In connection with these ideas, Vyacheslav Iosifovich Naidenov obtained the result that accounting for thermal processes in a pool with evaporation leads to an essentially nonlinear water balance equation. This approach, namely, bounded nonlinear effects, seems promising for constructing a forecast of fluctuations of the Caspian Sea level. To clarify the character of the nonlinearity of evaporation processes in shallow places, it would be suitable to account for multiple reflections of sun rays from the bottom and surface, which qualitatively explains the rapid heating of water in shallow places and the corresponding increase in evaporation. However, the quantitative aspects of this phenomenon have not been investigated and require serious study.

The surface of shallow waters is defined by the morphometry, i.e., by the dependence of the whole surface of the sea on the level. For the prognosis of global climate changes, it is necessary to clarify the reasons and dynamics of such changes in preceding climatic epochs. From the disposition of terraces on a coast line, paleoclimatologists and hydrologists try to reconstruct the humidity epochs of the paleoclimate. Such investigations have been conducted, for example, for Lake Victoria (Geits, Diesendorf) and partially for the Caspian Sea. Taking this into account, it seems very important to consider problems of the interaction of the sea and the coast for predictions of the level of self-contained pools (such as the Caspian Sea) because previous situations change its morphometry, which then influences future behavior. As already mentioned, marks relevant to a sea level at which the surface of shallow waters attains a local maximum should be stable. On the other hand, when the sea stays at the same level for a long time, it results in terracing of coasts. Therefore, these two processes should magnify each other. For the prognosis of level fluctuations, it is obviously important to construct a model of long-term behavior of the level of a self-contained pool in which, apart from the water balance equation, changes of the morphometry of the lake and its connection with the level behavior are taken into account.

The Fallaciousness of Linear Models

Since the mid-1970s, an unfortunate tendency toward consolidation of scientific judgments arose in hydrology in the USSR. All the ideas and approaches listed above came to be considered antiscientific, and only stochastic water balance methods were declared scientific. The first works on predicting the Caspian Sea level by stochastic methods were those of S. N. Kritskii and M. F. Menkel, who supposed that the dynamics of the Caspian Sea level is described by a linear difference water balance equation. The river discharges and visible evaporation (evaporation minus precipitation per unit of water surface) were considered the basic elements of the balance, and it was supposed that they are stationary stochastic processes. A linear dependence of the sea surface on the sea level was postulated for the whole scale of depths considered.

This procedure was later developed by employees of the Institute of Water Problems of the USSR Academy of Sciences and was introduced into practice by the Institute Soyuzgiprovodkhoz and the Ministry of Water Works of the USSR. This procedure was the base for substantiating the projects of diverting northern rivers into the Caspian Sea because the prognoses of fluctuations of the Caspian Sea level constructed by this procedure predicted a steady fall of the Caspian Sea level to –30 m B.S. by 1990.

The essence of the water balance prognosis developed at the scientific institutes of the Ministry of Water Works reduces to the following. After its coefficients are averaged, the water balance equation becomes a first-order linear difference equation with constant coefficients. Obviously, the solution of such an equation tends to a stationary value, which (not quite successfully) is called the "level of attraction". The free term of this equation corresponds to the influx to the Caspian Sea, basically from the Volga. If we now reduce this free term by the constantly increasing volume of water taken from the Volga for irrigation, the level of attraction will decrease, and we obtain a prognosis of a constant falling of the Caspian Sea level. In accordance with the traditions of the Ministry of Water works, a simple commonplace idea was camouflaged as a complex, highly scientific theory. To make the surroundings mathematical, the theory of equations with stationary stochastic coefficients, the rhetoric of the Kolmogorov extrapolation theory of stationary stochastic processes, estimation of variance, and so on, were used.

We emphasize that the Caspian Sea level has increased since 1979, contrary to the prognoses of its falling. This fact did not confuse the apologists of river diversion; they said it was only a random fluctuation, after which a phase of falling of the level will again begin.

We immediately note that the usage of the theory of equations with stationary coefficients in this procedure contained the crudest mathematical errors. One error is the supposition that the expectation of a solution to an equation with random coefficients is the solution of the equation with averaged coefficients. This would be true if the expectation of a product were equal to the product of the expectations of the factors. For people far from higher mathematics, we note that this is about the same as to say that the square of a sum is equal to the sum of the squares. In the materials of the commission of Academician Aleksandr Leonidovich Yanshin that was organized to discuss the problem of river diversion, Zelikin gave an example where the solution of the equation with averaged coefficients remains constant while the expectation of the genuine solution increases without limit. The assumptions about the stationarity of influx and evaporation are too rough and do not correspond to real processes; using the simplified linear model is unwarrantable. We add that visible evaporation defies direct measurement and is therefore not measured in the considered methodology but calculated only as a closing term of the linear water balance equation. In effect, it plays the role of a fitting parameter that forces the water balance equation to be fulfilled automatically.

Zelikina, Zelikin, and his postgraduate student from Germany Jorg Schulze organized a major series of computational experiments. The matter is that the diversion experts had a standard answer for any discrepancy between the real behavior of the sea and their prognosis: "That was an event of low probability". The considered forecasting technique was applied to alternating years of the known history of fluctuations of the Caspian Sea level. That is, based on data for previous years, a prognosis was calculated for the sea level; based on the prognosis considered, the probability of the events that actually occured in the subsequent years was then calculated. The outcome was startling. In almost half the cases, the probability was extremely small: frequently several percent, and sometimes less than one chance in ten thousand. It was found that improbable events should happen constantly: then why are they said to be "improbable"!?

Inference

It seems that all the approaches described above are suitable for modeling. Each has a grain of reason, and each captures one of the characteristic features of the process. But to join all these approaches in a unified theory would lead to a very complicated model, which could hardly be treated decently. Furthermore, there would be no assurance that all essential components of the phenomenon in question were taken into account. The impression created is that the Caspian Sea is comparable in complexity to a living organism.

This does not mean that prognosticating its behavior is impossible in principle. Indeed, it is sometimes possible to make reasonably certain predictions of responses to one or another situation not only of living creatures but even of human beings. It is only necessary to realize the level of relativity of such a prognosis, but not in the sense of probability theory as is frequently attempted. Probability theory finally describes only properties of a symmetry, which is absent here. The relativity should be understood in the sense that we certainly have an incomplete knowledge of the causes exerting a primary influence on the process considered. Therefore, the outcome is not determined inevitably as it would be by a certain firm law of nature.

It is necessary to be very cautious with scientific prognoses of natural phenomena. Even such a great physicist and mathematician as Christiaan Huygens (1629–1695) appeared in a very awkward position as a result of his prediction. He asserted that it is "scientifically proved that the surface of Jupiter can be sown with hemp, which is so necessary for the needs of the Dutch marine". Huygens was guided by astrological reasons: you see, Jove was considered the patron of mariners. The prediction of Huygens can be considered crackpot, but it has brought no harm to Jupiter, nor to mankind, nor even to the Dutch marine. It is quite another matter to predict fluctuations of the Caspian Sea level.

It was known very long ago that the Caspian Sea level is subject to strong unpredictable changes, and people tried to adapt to its changes. In the Middle Ages, the Shah found a severe, but simple, solution. He defined a flood-danger zone along the coast of the Caspian Sea and decreed the decapitation of anyone attempting to build a house there. Probably, he perfectly understood that human greed, rushing to seize convenient unoccupied land, can result in a tragedy. Another example is the walls of the city Derbent, which begin in the mountains, descend to the coast of the Caspian Sea, and continue rather far under the sea level. Thus, when the sea level varied (no difference whether up or down), the city remained enclosed by a wall that reached the coast of the sea.

In our "enlightened" days, the economic exploitation of its drained coast began as soon as the sea receded. Residences, industrial buildings, resort facilities, and sanatoriums were erected, roads and other communications were built, oil storages were constructed. All this construction was based on the "scientifically justified" prognoses of further lowering of the sea level. The rise of the level, which followed afterward, resulted in huge losses. The Ministry of Water Works had a vested interest in defending the validity of the prognoses of the Caspian Sea level falling because recognizing the error of these prognoses would mean an admission that exactly this ministry caused considerable losses to the country.

7.3 Mathematical Simulation in the Civil Engineering Design of the Leningrad Dam

The Leningrad dam construction project is an example of using inadequate models to masquerade as high science and simultaneously gives us an occasion to discuss the problems of the technological simulation of natural processes. In our judgment, there are essential discrepancies between the simulation of processes in engineering and physics and the simulation of widespread natural processes. The first obvious discrepancy is that natural processes cannot be repeated as many times as is required to check a model or a hypothesis. Moreover, if experimenters, when constructing a model in physics, can conduct an experiment in the conditions necessary for testing a hypothesis, in the case of natural processes, the investigators are forced to deal only with the set of measurements put at our disposal by nature. Therefore, even detecting the causal connections in natural processes presents methodological difficulties.

Another discrepancy between the simulation of physical processes and the simulation of natural processes is that the natural processes proceed very slowly. The study of natural processes is perhaps similar to genetics, where observations of genetic regularities are also hampered by the slow growth of organisms. We can give an example of a simulation based on near-term observations that resulted in an inadequate exposition of nature. In the region of the city Krasnodar, a reservoir was designed and constructed, its bed being above the profile of the landscape. That was achieved by building up the banks of the reservoir. It was so constructed that the water level in the reservoir was above the ground level in Krasnodar. The standard calculation methods based on extensive experience in constructing river dams indicated that the exfiltration of water under the banks of the reservoir should not be significant and should affect only a few kilometers from the reservoir. Years passed. After 25 years, water from the reservoir was detected in cellars in Krasnodar. The existing models could not predict this.

Neglecting these discrepancies when simulating large-scale natural processes can lead to errors, as is well demonstrated by the example of the Leningrad dam project. Construction of the Leningrad dam was lobbied by many influential organizations and departments of the former USSR. We participated in the environmental evaluation of the project. The chairman of the commission of the USSR Academy of Sciences for the environmental evaluation of the Leningrad dam project was Alexei Vladimirovich Yablokov; A. S. Mishchenko was the vice-chairman and Zelikin was a member of this commission. Because our activity was always considered a personal public service and because we spent only personal money, the organizations to which we appealed met our wishes and performed unique and extremely expensive research free of charge. We obtained not only aerial photography of the gulf but also satellite imagery with consequent decoding (the imagery was produced simultaneously in several different wave bands). Invaluable help in organizing these experiments was rendered by the cosmonaut Georgii Mikhailovich Grechko.

The Leningrad dam, as stipulated by the project, should be a system of dams extending from the north to the south in the region of the island of Kotlin (the city of Kronstadt), separating Neva Bay from the Gulf of Finland. Gates in the dams were designed to permit the passage of the Neva water. They were to be closed at the approach of a Big Wave. The designers stated that the width of these gates would be quite sufficient to ensure a normal drain of the Neva in the usual situation. For proof, they constructed an experimental basin about the size of a huge hall, patterning the geometric form of Neva Bay. About five million rubles were spent for this building. In the experiments, all was obtained safely.

A Little Bit of History

The river Neva is a unique essential water source in this region. It empties into the Baltic Sea in the eastern part of Neva Bay through an extensively branching delta network. The water of the Neva flow further along the whole of Neva Bay from east to west and through the northern and southern gates, where the current is divided by the Kotlin Island. It then meets the Gulf of Finland and intermixes with the salt water.

St. Petersburg, like a horseshoe, envelops the whole of Neva Bay and even the eastern part of the Gulf of Finland. St. Petersburg is a wonderful city, but one of its troubles since immemorial times is the lack of a single wastewater treatment facility. Its wastewater has entered the Neva directly. It is necessary to add that the Neva is already not as pure as it was long ago; it carries discharges from polluted Lake Ladoga, fertilizers washed from fields with forced irrigation, etc. And Neva Bay is a very small bay with a feeble current. It would have turned into a refuse pool long ago if not for the floods.

The coast of Neva Bay is subject to regular autumn-winter seasonal floods. These floods are due to large solitary waves that arise in the Baltic Sea. Reaching narrow Neva Bay, they raise the level sharply. The level of flooding of the city in different years fluctuates from 1.6 up to 4 meters. Leaving, the Big Wave carried away to the sea all ground accumulations, periodically stripping Neva Bay and the Gulf of Finland of dirt. Certainly, the city suffered. There were very strong floods that remained in the national memory as disasters. A famous poem by A. S. Push-kin, *The copper rider*, described one such flood; between the lines, one can read speculations about the human right to interfere with natural processes.

Different projects to guard the city from floods have been developed since the times of Peter the Great. The most popular were projects patterned on those of Northern European countries, in the form of countercurrent dams. None of these projects was embodied before now.

With the blessing of the city chief, Politburo Member of the Central Com-mittee of the Soviet Union Communist Party, Grigorii Vasilievich Romanov, an intense building of suburbs in places most subject to floods was started. The city continued to grow in the seaward direction. Many microregions were created on platforms won from the sea by filling with ground. Thus, at the center of the large megalopolis, there exists a practically self-contained pool, Neva Bay with a surface of 329 km^2, and the well-being of the Petersburgians wholly depends on its ecological status.

Designing the Leningrad dam began in 1966. In the technical assignment for compiling the report of the technological-economic basis of the project, the follow-ing was specified: (1) to protect Leningrad from floods on the part of the Gulf of Finland, even once-in-1000-year floods; (2) to protect coastal regions of the Neva from winter over-ice floods; and (3) to create conditions ensuring the sanitary status (with respect to the standards) of the Neva Bay water area and branches of the Neva delta within the city limits. It was planned to consider four variants of protective constructions: two in the west and two in the east.

In 1968, the State Committee on Economic Construction of the USSR, contrary to the plans, proposed to consider only one western variant with separation of Neva Bay from the Gulf of Finland and disposition of protective constructions along the line Lisii Nos–Kronstadt–Lomonosov. All remaining projects and variants were discarded. Despite its many negative points, the authorities began to push only this variant with persistence worthy of a better cause. These negative points were sharply aggravated as a result of an urgent rush to finish construction, which met public resistance.

In a preliminary expert evaluation of the project, the following requirement was noted: "At separation of Neva Bay from the Gulf of Finland by protective constructions, their building should be preceded by the realization of measures to neutralize, disinfect, and clean wastewater emptied into its water-collecting basin". But the State Committee on Economic Construction of the USSR proposed constructing treatment facilities not before but only on an overtaking schedule.

Construction of the dam began in 1979, and construction of treatment facilities not only did not precede but also lagged behind the construction of the dam. In addition, the construction was conducted with violations of the engineering project. The most substantial violation was that in December 1984, the northern passage (Lisii Nos–Kronstadt) was closed by a blind technological dam without the planned gates. This resulted in a critical ecological situation in Neva Bay. It is necessary to note that at the beginning of construction work, the scientific justification of the selected design was simply lacking.

Errors in Simulation

The complex ecological justification and impact evaluation of building of protective constructions was missing in the project. The role of such an impact statement was played by reasonings based on hydraulic and mathematical simulation invoked to prove that the protective constructions will exert little influence on the field of currents in the separated part of Neva Bay. To analyze the influence of the protective constructions on the hydrodynamic field of currents in Neva Bay in detail and to estimate more precisely the influence of the constructions on environmental circumstances, a general hydraulic model was especially created in the Vedeneev All-Union Research and Development Institute of Hydraulic Engineering.

Results of the research on this model were widely used to justify the assertion that the protective constructions will have practically no influence on environmental circumstances, because the change in the average field of currents in Neva Bay would be slight and rather localized. Actually, the environmental circumstances are defined by fine biological and biochemical processes taking place in the water, which depend not on the average flow characteristics of the pool but on details of the velocity distribution determining the salinity of water, its temperature, transparency, oxygen saturation, etc., in the considered case and also on the density of chemical admixtures and contaminations.

Our analysis during environmental evaluation showed, in particular, that the design of the general hydraulic model used to simulate the influence of the protective constructions on the hydrodynamics of Neva Bay contained essential deviations from the principles of hydraulic simulation and that the necessary methodological research was not performed.

The main idea in the hydraulic simulation of the mouths of big rivers is to pass from the study of a three-dimensional current to a study of a two-dimensional current describing the depth-averaged behavior of the modeled current. From the mathematical standpoint, we must average the equations describing a turbulent flow of an incompressible liquid. The set of equations describing the planar field of depth-averaged velocities of a riverbed stream depends on four dimensionless parameters: the Struchal, Frud, and Reynolds numbers and a friction parameter. The possibility of simulating large-scale flows on small-scale models is based on the fact that currents, both in nature and in the model, are driven by the same set of equations. Therefore, if the dimensionless parameters figuring in a set of equations and boundary conditions coincide in magnitude for the model and for nature, then its dimensionless fields of velocities in dimensionless coordinates will coincide.

The simulation of large-scale flows on geometrically similar models, i.e., with proportional contractions of planar (b) and vertical (h) sizes, can meet engineering difficulties. For example, for the problem of the current in the mouth of the Neva, we have $h = 3$ m and $b = 30$ km. With these conditions, even for a model with a planar size of 100 m, the depth of the stream should be only 1cm. It is clear that to work with a stream with a 1 cm depth would be very difficult and to build such a model would be practically impossible.

One way out of these difficulties is to simulate on models with disproportionate vertical and planar scales such that the contraction is less in the vertical than in the planar dimension. It is known that it is generally impossible to achieve coincidence of all the dimensionless parameters for the model and for nature; therefore, strictly speaking, the precise engineering simulation of riverbed streams is impossible. Nevertheless, simulation appears to be useful in some cases. This circumstance is conditioned by the fact that for a class of currents, the field of currents may be independent of the specific value of one (or several) of the parameters in a particular range of values of the dimensionless parameters (in a so-called automodel area of the relevant parameter). Then, instead requiring coincidence of the dimensionless parameters for the model and for nature, it is sufficient to require that the values of the relevant dimensionless parameters belong to the automodel area.

Rich experience in simulating riverbed streams on models with disproportionate scales indicates that this disproportion must not be large (no more than five times). With large scale disproportions, special methodological research is required in a wide gamut of current conditions to determine which parameters can be modeled on such constructions. Moreover, for each specific class of currents, it is necessary to analyze the boundaries of the automodel areas (on the basis of model experiments and measurements in nature) and the possibility of realizing current conditions for which the characteristic dimensionless parameters lie in the automodel area. This operation, which is necessary to justify using the simulation, was not done in planning the general hydraulic model in this case. In fact, the scale disproportion ratio $k = 10$ (instead of $k < 5$ as recommended in the literature) was accepted without any justification, the automodel condition for the Reynolds number was not determined, and the boundaries of the automodel area for the friction parameter were not defined. Because of these essential shortcomings, the model has appeared practically unsuitable for simulating currents in Neva Bay, and the experimental data obtained on it should not be used to confirm or disprove any conclusions concerning the influence of the protective constructions on the hydrodynamics of the separated part of Neva Bay.

This conclusion becomes even more obvious if we recall that the hydraulic models basically cannot model wind currents, which can essentially influence currents in Neva Bay (for example, it is known that the influence of wind on currents in Neva Bay is essential in 90 cases out of 100). Currents in shallow waters are modeled with the least reliability on hydraulic models, but the basic biological and biochemical processes determining the water quality proceed in exactly these places.

Environmental problems deserve special considerations. The processes determining environmental circumstances in Neva Bay depend not only on the average field of velocities but also on a series of fine hydrodynamic effects, which are not modeled by hydraulic models. This applies equally to stratified currents defining salinity levels, to wave processes intermixing water and carrying out contaminants, to settling of particles born by the river, etc.

The results of hydraulic simulation on the general hydraulic model were widely used by the designers and builders to introduce the idea into the public consciousness that the influence of the protective constructions on environmental circumstances in the region would be insignificant. The reasons mentioned above show that it is impossible to recognize such conclusions as justified. The limitations of hydraulic simulation for estimating environmental problems and the shortcomings of specific models were well known to its developers. The construction of this largest model in the world (it cost about five million rubles) can serve as a brilliant illustration of methods used to act on public awareness.

Comparison with Nature

One could certainly consider all these arguments inessential. But now, when construction of the dam is practically finished, there is no necessity to test theoretical arguments. It is enough to compare the environmental situation in the region of Neva Bay before and after construction of the dam. There is a possibility to evaluate the influence of the protective constructions on natural and anthropogenic processes by comparing real facts. For this, we have considered two stages of ecosystem functioning: first, prior to the beginning of closing the northern passage of Neva Bay by the dam, i.e., to 1983 inclusively; second, after this closing. Since 1984, in connection with construction of the dam, there have been sharp changes in practically all elements of the ecosystem of Neva Bay and the eastern part of the Gulf of Finland. The fluctuations of the level of Neva Bay have changed (phase delay of oscillations, reduction of an amplification coefficient, suppression of oscillations in the high-frequency part of the spectrum).

Salinity

The inflow of salt water into Neva Bay from the Gulf of Finland was sharply reduced. It is doubtless that as a result of constructing the dam, the maximum salinity values of all the waters of Neva Bay (except for the ship channel) now do not exceed the salinity of the Neva water (0.08 g/l), while before constructing the dam, the maximum salinity values exceeded the salinity of the Neva water at some times in practically all of Neva Bay including even its near-coastal part. It is clear that this phenomenon could not be predicted on the hydraulic model because this model was planar: it featured depth-averaged current. The infiltration of salt water into Neva Bay resulted from a near-bottom countercurrent. The freshwater current from the Neva went on the surface, and the heavier salt water went below and comprised a compensating countercurrent. Because the width of the dam gates was much less than the former width of the river, the countercurrent was almost completely stopped, and this resulted in the drop in salinity. For this reason, after constructing the dam, even the salinity in the region of the ship channel is hardly above that in the shallow parts of Neva Bay.

Before 1985, even in the coastal zone, the maximum salinity values reached 1 g/l, whereas after 1985, there was a drop of salinity down to 0.08 g/l, i.e., a more than ten times reduction. Before 1985, even the annual average salinity values in the first three alignments, considered from the mouth of the Neva (mouth, Lahta–Strelnya, Lisii Nos–Petrodvorets) achieved magnitudes of 0.15 0.35 g/l. After 1985, these densities were stabilized at the level of salinity of the Neva water (0.08 g/l). This means that previously (before constructing the dam), the salt water from the Gulf of Finland penetrated all parts of Neva Bay. Such infiltration of salt water was now stopped. Before 1984, in the western part of Neva Bay adjacent to the dam, the maximum salinity values, varying from 3 up to 5 g/l, were comparable to the salinity values of waters in the eastern part of the Gulf of Finland near the dam. After 1985, in Neva Bay adjacent to the dam, the maximum salinity steadily reduced to 1 g/l, while the salinity remained practically unchanged in the eastern part of the Gulf of Finland adjacent to the dam.

Therefore, as a result of constructing the dam, desalinization of the water in the western part of Neva Bay has occurred, and the salinity of all the water in all parts of Neva Bay was considerably reduced. The changes in salinity are also confirmed by the vanishing of saltwater species of plankton. Because the salt water reached the eastern part of Neva Bay only as a result of hydrodynamic currents, the changes in salinity irrefutably prove that the hydrodynamics of the separated part of the bay has changed.

The same shows that the correlation between salinity changes in the eastern part of the Gulf of Finland and in Neva Bay disappeared after constructing the dam. After 1985, practically all hydrochemical density indexes increased (oxygen decreased) in the southern coastal strip as compared with average densities in Neva Bay, i.e., there was a redistribution of contaminates. This fact can only be explained by changes in the hydrologic conditions in Neva Bay connected with constructing the dam. To the present, the stability of outflowing transit streams in the bay has increased, and stable slow-current and stagnant areas, where pollutants began to accumulate, were formed near the coasts and in front of closed sites of the dam. According to research by the State Hydrologic Institute, new, previously uncharacteristic regions with alternating zones of time-lagged water exchange and jet flows were formed in a five-kilometer zone on both sides of the dam.

Contaminations
On both sides of the southern passage, the nitrite density now exceeds the level of Allowed Marginal Densities (AMD) by two or three times. In this previously most productive region, the quality of bottom grounds has sharply declined, mass diseases of zooplankton are observed, and contamination by fecal microorganisms has increased. According to data from the beginning of the 1980s, the zone of greatest transparency (1–2 m) enveloped the central part of the pool and occupied about 60% of the water area.

The second zone with a minimum transparency (0.05–0.5 m) extended along the southern coast of the bay. Along the northern coast, the transparency varied from 1 up to 1.5 m. Comparing the data from the beginning of the 1980s with data from the beginning of the 20th century shows that the observed transparency patterns were also characteristic for the beginning of the century. Analysis of 1983–89 satellite imagery shows changes in water quality classes of Neva Bay with a tendency to a sharp increase of areas occupied by waters with strong and very strong contamination (water quality classes 4 and 5).

According to the data, the area occupied by water of the first contamination class (conditionally pure) has decreased from 80% in 1983 to 13% in 1986 and 10% in 1989. The area occupied by water of the fourth and fifth contamination classes has increased from 0.2% in 1983 to 14% in 1986 and 28% in 1989. According to satellite imagery, zones of strong contamination are concentrated along the southern and northern coasts and in front of the dam. We emphasize that the contamination zones were localized in 1983 and did not threaten quality changes in the entire bay.

With the beginning of the dam construction, the water contamination situation changed sharply: water of quality classes 2, 3, and 4 now dominate (on an area of 80% with a preponderance of class 3). A new hydrologic phenomenon was the appearance of large zones of turbidity in coastal parts and, in particular, west of the dam, where infiltration of river water of class 1 (conditionally pure) is practically unnoticeable. With continuation of the dam construction, there was a further increase in the contaminated area, reaching critical sizes of 90% of the eastern part of the Gulf of Finland. The dominance of water of quality classes 4 and 5 (strong and very strong contamination) became especially noticeable.

Analysis of the long-term observations shows that the area of water of class 1 (conditionally pure) shows a steady tendency to reduction: in six years, it was reduced eight times. Water of class 2 (slightly polluted) has slightly increased in area (1.5 times). The area of water of class 3 (moderately polluted) has increased more than two times. But the changes in area of waters of classes 4 and 5 (strong and very strong contamination) are especially great. Their area has increased almost 100 times! It now occupies more than half the area of the eastern part of the Gulf of Finland.

To determine zones of different wave disturbance, aerial radar images were obtained. Superposing maps of water quality classes and wave disturbance levels showed that the zone of reduced chop, as a rule, coincides with the zone of increased water contamination. All these changes resulted in acute modifications of the hydrobiological system of Neva Bay, caused a disastrous ecological and sanitary-hygienic situation in the bay and the eastern part of the Gulf of Finland, and shaped an extensive zone of heightened ecological risk within the limits of the urban agglomeration of St. Petersburg.

Flora

The part of the dam closing the passage north of Kotlin Island was built first. Because of hurry, the builder made a blind closure, not even leaving the planned gates; they were to be built later. Troubles came immediately. The bay became overgrown with blue-green algae. These algae are very insidious. To protect their ecological niche, they excrete substances similar in action to nerve and paralytic poisons. An epidemic of nerve diseases erupted in St. Petersburg; the authorities carefully kept their silence. Bathing in the bay was prohibited, but no one can stop children. Even after the gates were at last constructed, the ecological situation improved little.

Analysis of phytoplankton and periphyton communities as well as their structural and functional indices showed that in the 1980s, the phytoplankton, as before, was under the influence of biological sewage from Lake Ladoga and the river Neva, and the condition of Neva Bay was quite satisfactory. The development of its own phytoplankton that is typical only for eutrophic reservoirs was noticed only in stagnant sections of the littoral zone in the southern area.

In the western part of Neva Bay, the phytoplankton was formed at the expense of the plankton in the eastern part of the Gulf of Finland. The blue-green algae were of no importance in forming the plankton biomass before 1984. The main role in forming organic matter was played by diatoms in the spring and by chlorococci and yellow-green algae in the summer.

The factor limiting the development of phytoplankton before erecting the protective constructions was the high flow in Neva Bay that impeded mass production, as well as the low water transparency. The provision of phytoplankton with biogenic elements including mineral phosphorus (the main element of mineral nourishment determining algal development) had exceeded the phytoplankton demand and could not limit its development.

The N:P ratio was also sufficiently optimal for unlimited development of the phytoplankton complex, but massive phytoplankton development was not marked. Since 1984, however, some structural changes in the phytoplankton complex occurred increasing the proportion of colonial species of blue-green algae (by 1986, the changes spread over the entire water area of Neva Bay). In the process, the distribution character and level of algal development were extremely nonuniform within the bay water area. Maximum size values have been marked since 1983 in the southeast section of Neva Bay with its peak development in 1986 reaching 27.3 million cells/l with a background value of 2 to 7 thousand cells/l. The same tendencies of increasing the blue-green algae population appeared in the central transit section as well as in the bar area with maximum development in 1986.

The most essential changes occurred in the phytoplankton dynamics in Neva Bay. As the Northwest Administration of the State Hydrometeorological Committee showed, the difference between the phytoplankton populations in the gulf and in the bay reduced sharply immediately after the northern passage was closed. From 1984 to 1988, the total phytoplankton biomass increased in the open part of Neva Bay and decreased 2.3 times in the Gulf of Finland. The dam building caused essential changes in composition of phytoplankton species. Before the dam erection, the blue-green biomass in the water area of Neva Bay had varied within 0–20% of the total phytoplankton biomass, but after 1984, it has ranged from 60% to 97%. Evaluating possible factors responsible for such significant changes in the microphytocenosis of Neva Bay allows concluding that changes in the hydrodynamics of the water mass and the flow velocities in the central transit section are of priority because before the dam erection, enhanced hydrodynamic activity had been the main factor inhibiting such large-scaled production of blue-green algae in Neva Bay.

Analysis of the coastal aquatic vegetation indicates that in the early 1980s, the northern bank of Neva Bay was almost devoid of overgrowth in its northwest part with a smooth increase of overgrowth area toward Lisii Nos, where essential development of the coastal aquatic vegetation was marked. On the southern bank, the overgrowth formed small accumulations with a maximum in the Strelnya area. By the 1980s, the overgrowth area in the littoral zone of Neva Bay was about 6.5–7.0 km^2, that is, 2% of the total water area. The overgrowth area was about 3 km^2 on the southern bank and 2.5 km^2 on the northern bank. The dominant species were cattails and reeds. Filamentous green algae (fouling) could be found in all small coves and creeks with a maximum development in stagnant shallow waters. The factor limiting the development of the vegetable complex (fouling) in the early 1980s was the active hydrodynamics of the water mass.

In the late 1980s under conditions of slight change in the high level of mineral nourishment of plants, the overgrowth with filamentous algae was enhanced in the shallow water areas; this is evidence that the factor of drastic flow changes in Neva Bay resulting in the formation of vast areas of delayed water exchange noticed during 1984–87 most likely worked here, rather than an increased escape volume of poorly treated wastewater.

Fauna
Investigation of zooplankton in Neva Bay showed that at the beginning of the 20th century, its species content was formed mostly of species typical of Lake Ladoga. Before the 1980s, the number of species of rotifers, one of the most indicative groups of zooplankton, was unchanging in Neva Bay and Lake Ladoga. The zooplankton in Neva Bay originated from the Ladoga complex transformed during transfer by the Neva river, enriched with phytophilous species that developed in the floodlands of the south bank reservoirs of Neva Bay as well as with some saltwater species from the eastern part of the Gulf of Finland and some marine species able to function in conditions of strong desalting. Almost 50% of the total

zooplankton biomass was rotifers, 30% was infusorians, 20% was cladocerans, and about 3% was copepods. Before 1984, the rotifer proportion in the plankton in the central transit area as well as in the northern area was 75%, decreasing to 25% in the southern area.

After closing the northern passage with the dam, structural and functional changes occurred in the zooplankton association with a high-volume development of the rotifer complex typical of eutrophic waters. The first outbreak was marked in 1985. We note that in 1987, after the gate constructions had been cleaned and additional passages had been produced, a short-term reduction of the rotifer population was observed, but in 1988–89, the peak high-volume development of the rotifer zooplankton exceeded the background level 4 to 5 times.

The pouring-off of the southern part of the Leningrad dam began in 1986 and was finished in 1988, producing a blind barrier 3 km long. In 1988, the peak zooplankton development in the southwest part of the bay water area at the stations adjoining the dam exceeded the 1986–87 background indicators 9 to 10 times. The main component of the peak was rotifers.

Analysis of the taxonomic composition and the structural and functional indicators of the benthic fauna showed that by the beginning of the 1980s, a representative feature was the high-volume development of molluscs, reaching its peak (with a biomass up to 1 kg/m^2, which is the limit value known for fresh-water reservoirs) in the bar area. This complex of benthic animals played the major role in the biological self-cleaning of the water, where benton, not plankton, was the most important factor in the purification process.

The majority of the precipitable suspension carried by the Neva river was salvaged in the bar area of Neva Bay. The destruction rate of organic substances in the eastern part of the bay and the production amount were very high here, significantly exceeding the initial production in Neva Bay itself. This indicates that the allochthonous organic substances brought by the Neva played a deciding role in the ecosystem functioning, and this is a specific functional feature of the benthic zoocenosis. The main factor determining the composition and productivity of the benthic associations was the aeration level in the benthic water layers, that is, saturation of the bottom layers with oxygen and the trophic factor as a result of the active hydrodynamics of the water mass.

Data from the period after 1985 revealed some tendencies toward structural changes in the benthic zoocenosis as well as a tenfold reduction in its biomass in the bar area. The observed tendencies indicate that drastic changes for the worse in the total biological condition should be expected in the very near future in connection with disruption of the functioning of the main biological component of the self-purification of Neva Bay's benthic zoocenosis. According to recent research, the whole condition of Neva Bay can be described as unsatisfactory and even catastrophic in its sanitary and hygienic indicators: the coliform indices of 88.6% of the water samples do not satisfy GOST (Russian State Standard). In total, the water quality in the bay as evaluated by the coliform index is bad in 86.3% of the samples and unsatisfactory in 7.5% of the samples.

7.4 Mathematical Simulation of Physical Processes

As the understanding of the essence of a phenomenon was refined and the heart of the matter was approached, models usually moved further and further away from the initial, tentative, primitive constructions. The planetary model of the atom of Ernst Rutherford (1871–1937) played a productive role during the development of physics. But when the Heisenberg indeterminacy principle, which prohibits the simultaneous localization of the position and velocity of a particle, was finally recognized and when it was established that the hopping of an electron from orbit to orbit can happen only in a quantum manner, then the picture of electrons as balls revolving similarly to planets around the nucleus became an anachronism. Bohr's model (Niels Bohr (1885–1962)), which replaced Rutherford's model, lost an essential obviousness but gained much more in the sense of the possibilities of its usage.

The same occurred with the electron spin. Originally, it was represented as a gyration (although also quantized) of electron-balls about the axis (whence the term "spin"). But the electron is now quite often represented as a mass point, and gyration then loses its meaning. Therefore, the term "an interior degree of freedom", which is free from analogy with a rotating ball is more often used.

Modern physics moves further and further away from mechanistic obviousness and becomes more and more mathematicized. This sometimes provokes protest from physicists. In particular (according to Yakov Borisovich Zeldovich), William Thomson (Lord Kelvin, 1824–1907) offered a rather eccentric unit of measurement for the degree of certainty of physical and mathematical formulas: he proposed measuring the inverse degree of certainty in inches. Namely, short formulas are much more likely true, and a formula more than two inches long is certainly false. This is interesting. But what do we do with the lengthy intermediate formulas, which arise when deducing the short formulas? If we find them to be false, will the final short result then be false too? Or, on the other hand, if we arrange some short formulas into one long one, will their truth be destroyed?

Arkadii Binusovich Migdal brought Thomson's thesis to its logical conclusion by offering a definition: "Physics is the art of bypassing mathematics". Perfect! However, we can ask how to calculate the results of planned experiments (the main goal of physics in the opinion of many physicists) without using mathematics. Or do we allow the use of elementary arithmetic, which is excluded by definition from the domain of mathematics? There is one more difficulty. Elementary particle physics frequently deals with energy values that are not yet experimentally accessible. In particular, such a considerable and vigorously developing area of elementary particle physics as chromodynamics till now almost lacks support from direct experimental proofs. Such a support needs particles possessing an energy inaccessible on modern accelerators. Another example is astrophysics and cosmology, where it is difficult to imagine any physical experiments at all even in the very long-term future. What will play the role of the validity measure in physics with a complete lack of both mathematical and experimental proofs? It seems likely that the role of mathematical models in such questions is unique.

A striking example of simulation is the twistor theory of Roger Penrose. Penrose, being Platonic, believed in the substantial existence of an eternal world of ideas. Natural laws are perfect in this world of ideas, but they become less perfect when reflected in our physical world. According to relativity theory, time stops for an object moving with the speed of light, and in some sense, one obtains eternity (in theology, Light is the prerogative and the symbol of God!). Therefore, natural laws must be described in systems moving not with ordinary velocities but with the speed of light. To realize this program (the so-called twistor formalism), Penrose had to complexify the Minkowski space-time and invent its map to a light cone, which began to be called the correspondence of Penrose. As a result, several fundamental equations in mathematical physics, such as the Maxwell equations, the massless Dirac equation, and the linearized Einstein equation, when written in the twistor formalism, became really perfect: arbitrary holomorphic sections of relevant bundles appear as their solutions.

But the twistor formalism is difficult to combine with the quantum approach. And the problems of quantum field theory recently claimed a leading interest both in physics and in mathematics. In quantum mechanical models, the basic role is played by the Heisenberg indeterminacy principle treated as commutation relations between the basic quantum mechanical operators. The quantization process in mathematical constructions even came to be understood as simulation of commutation relations (one more interpretation of the term "simulation"). There is also a change of the scene in which the game is played. The role of classical space-time, the domain of functions representing the status of a quantum system, is now played by infinite-dimensional manifolds: the set of Riemann surfaces (in string theory), the Virasoro algebra, and so on.

The remarkable physicist Richard Feynman remembered why he became a physicist. In childhood, he was shocked by a commonly known unpretentious physical experience: if amber is rubbed by wool, the amber is electrified and begins to attract small objects. Feynman decided to become a physicist to understand why this happens. He writes in his memoirs that he graduated from the university, studied physics, made many widely known physical discoveries, won the Nobel Prize, but still cannot explain why amber is electrified when rubbed by wool. And this is not a unique problem with no adequate mathematical model for its explanation.

Paradoxical as it may seem, the physics of rigid bodies still cannot explain why rigid bodies are rigid. The distances between the atoms in comparison with the sizes of the atoms themselves are as large as the distances between the stars in galaxies in comparison with the sizes of the stars. It would seem that nothing prevents one rigid body from infiltrating another!

One of the most difficult problems that arouse the acute interest of mathematicians and physicists is the problem of phase transitions. How do transitions from one state of a substance into another proceed, i.e., how are the processes of melting, crystallization, and so on described in quantum language? To clarify these problems, many different models were created. One of the most beautiful is the Ising model. To calculate the quantum effects responsible for phase transitions, an infinite discrete lattice on a straight line or on a surface is considered. From the physical standpoint, if a two-dimensional torus is taken as a surface, then a periodic lattice is considered. Each site of the lattice is in this or that quantum state. It is assumed that each site influences only adjacent sites. This assumption generates a dynamic system with a finite or an infinite number of variables. The problem is to find the asymptotic behavior of solutions of this system. Recently, A. S. Mishchenko and his coauthors succeeded in expressing a generating function of the Ising model in terms that depend not on the specific lattice but on topological invariants of the surface. The topological invariants are mathematical magnitudes that are constant under any continuous variation of the functions defining the model and the initial conditions. If our conclusions are based on topological invariants, this assures us that inaccuracies in the input data of a model do not influence the outcome. In the case of the Ising model, the so-called *Arf*-invariant was used (an invariant of some quadratic form over the field Z_2, its peculiarity being the nonlinearity in terms of the initial parameters of a problem.

The most mysterious phase transitions are to superfluidity and to superconductivity. Even at the dawn of superconductivity theory, the idea of an energy gap was advanced. The energy spectrum of the relevant quantum system is divided by an empty interval, an energy gap. The eigenvalues of the superconducting state are above this gap; eigenvalues of the so-called ground state, the rest state, are below. To pass from a superconducting state to the ground state, a rather essential energy decrement is needed. Therefore, rather small power losses at casual collisions of particles, i.e., the processes generating resistance and reducing a current in the usual conditions, do not influence a current in the case of a superconducting state. Victor Pavlovich Maslov and A. S. Mishchenko recently investigated the asymptotic behavior of the energy spectrum. The research results specify the possibility of the existence of an energy gap for not only superlow but also high temperatures. To realize high-temperature superconductivity, it is necessary to understand how to translate a substance into a state above this energy gap. Physicists are now struggling with this problem almost blindly, trying to find physical conditions and power fields, selecting materials at random, and so on. The lack of precise mathematical models, which could facilitate this search, is clearly felt.

Figures of Equilibrium of a Rotating Liquid

Excellent patterns of modeling were given by Isaac Newton (1643–1727). For instance, the contraction coefficient of the Earth was calculated in *Principia*, Book III. Newton used the simplest model to describe the equilibrium figure of a rotating homogeneous gravitating liquid. He imagined two boreholes with unit cross sections, one from a point on the equator to the center of the Earth and the other from the North Pole to the center. Equating the "weights" of liquid in both holes and taking centrifugal forces into account, Newton showed that the Earth's contraction near the poles, that is, the ratio of the difference between the equatorial and the polar radii to the mean radius of the Earth, equals 1/230. This assertion contradicted the views of many astronomers of that time belonging to the school of Gian Domenico Cassini (1625–1712). They reasoned that the Earth is extended rather than flattened near the poles.

There was a long discussion, and finally the geodesic measurements made in Lapland by Pierre Louis Moreau de Maupertuis (1698–1759) and Alexis Claude Clairaut (1713–1765) fully confirmed Newton's calculations. Voltaire made an epigram on his friend Maupertuis; it can be considered a humorous comparison of speculative and empirical epistemology.

Vous avez confirm dans les lieux pleins d'ennui
Ce que Newton connut sans sortir de chez lui.

We recall a phrase of Confucius relating to a wider problem than the natural sciences, namely, to understanding the essence of the Universe: "Three ways lead to wisdom: the way of imitation, the simplest; the way of experience, the hardest; the way of meditation, the noblest". Using Newton's idea, Colin Maclaurin (1698–1746) proved that the ellipsoid of rotation with the contraction calculated by Newton's method can indeed serve as an equilibrium figure. It is striking that the modern data on the Earth contraction (1/290) is not so different from Newton's result; the difference is explained by the heterogeneity of the Earth.

A real scientific sensation was the work of Carl Gustav Jacob Jacobi (1804–1851), who found that triaxial ellipsoids without rotational symmetry can serve as equilibrium figures. The next success was achieved by Poincaré and Aleksandr M. Lyapunov (1857–1918). They found nonellipsoidal (so-called pearlike) equilibrium figures. Poincaré considered these figures stable, but Lyapunov proved that they are unstable. James Hopwood Jeans (1877–1946) later suggested a conjecture that the Moon was formed by tearing off from the Earth as a result of the evolution of an unstable pearlike figure.

All the aforementioned authors considered a gravitated liquid rotating as a solid body. Consideration of equilibrium figures with internal flows began with the works of Bernhard Riemann (1826–1866), Peter Gustav Lejeune Dirichlet (1805–1859), and Richard Dedekind (1831–1916). It is noteworthy that almost all outstanding mathematicians of the 19th and the beginning of the 20th century (Gauss, d'Alembert, Laplace, Lagrange, Legendre, Liouville, Chebyshev, Cartan, Schwarzschild, Volterra, Steklov, to mention only a few) contributed to the problem of equilibrium figures and their stability because of the indisputable importance of these problems in astronomy and cosmology.

All this story gives us the opprtunity to discuss the radical difference in the views of Poincaré and Lyapunov on problems of mathematical modeling. Poincaré, who obtained his results using not so rigorous reasoning and often by simple analogy, wrote, "It is possible to make many objections, but the same rigor as in pure analysis is not demanded in mechanics". Lyapunov had an entirely different position: "It is impermissible to use doubtful reasonings when solving a problem in mechanics or physics (it is the same) if it is formulated quite definitely as a mathematical one. In that case, it becomes a problem of pure analysis and must be treated as such". Bowing to the genius of Poincaré, we unconditionally take the side of Lyapunov in this debate.

We recall the heated discussions of the mathematical models that proved the inevitable falling of the Caspian Sea level, the models that justified the projects of diverting northern and Siberian rivers to the south. At that time, some mathematicians spoke condescendingly of the scientists who elaborated these models: "They are engineers; one cannot demand rigor from them". In some sense, this is true. It is not necessary that a result in some applied science be realized as a strict mathematical therem. But it was not a question of mathematical purism. We spoke only of reliability. Our main objection against predictions of falling of the Caspian Sea level was that this model was inadequate. We did not demand taking all the factors into account. We simply asserted that main causes were omitted. It was clear that a linear model and only probabilistic factors were obviously insufficient for explaining the real behavior of the level. The mathematical fallaciousness was only an additional argument showing the scientific inconsistency of the authors: if they reveal ignorance of the mathematics with which they wish to cloak their reasoning, then there are no serious reasons to trust their professional arguments, especially as these professional arguments were not so complicated and it was easy to find weighty counterarguments.

Nowadays, using mathematical models to justify statements in the natural sciences, technology, and engineering has become fashionable, but this process sometimes takes almost sick forms. The point is that mathematics has the power to ennoble and even to sanctify scientific arguments. This property of mathematics sometimes turns into something quite different, into a not-so-honest trick, almost deceit. In our practice of evaluating the impact of environmentally dangerous national-economic projects, we more than once encountered a situation where cachectic, pseudoscientific concepts put on the mask of high science by using mathematical models. The authors of such manipulations try to use the most uncommon mathematical constructions, hoping that experts with a lack of understanding would neglect their duty and consider the corresponding statement "high science".

We object to attempts to turn modeling into an independent branch of science. It seems to us that this would not lead to discovering any new statements unfamiliar to really working scientists. The result would be the flourishing of pseudoscientists having little if any acquaintance both with mathematics and with a specific object of modeling.

It seems that the only effective method for designing models that was discovered by Mankind is the professional mathematical treatment of real facts and intuitive approaches that are related to the phenomenon in question. In other words, in a more figurative style, we should seek an audience with Her Majesty the Queen and Servant of All Sciences, Mathematics, and beg Her Majesty most humbly on behalf of the corresponding concrete Science for generous help: to serve in the creation of a new, important, and beautiful Model!

References

Liapounoff, A. M., 1909, "Sur une classe de figures d'équilibre d'un liquide en rotation, *Annales de l'École Normale*, 3rd series, 26: 473–483

Mishchenko, A. S., 1992, "Stationary solutions of nonlinear stochastic equations", in: *Global Analysis. Studies and Applications* (Borisovich, Yu. G. and Glicklich, Yu. E., eds.), Berlin, Springer-Verlag: 217–236

Mishchenko, A. S., Dolbilin, N. P., Zinov'ev, Ju. M., Shtan'ko, A. M., and Shtogrin, M. I., 1996, "Gomologicheskie svojstva dimernyh pokrytij reshetok na poverhnostjah", *Func. an. i ego prilogenia*, 30(3): 19–33 (English translation: "Homology properties of dimer coverings of lattices on surfaces", *Functional Analysis and its Applications*, 30(3), 1997: 163–173.

Mishchenko, A. S., Dolbilin, N. P., Zinov'ev, Ju. M., Shtan'ko, A. M., and Shtogrin, M. I., 1999, "The two-dimensional Ising model and the Catz-Word determinant", *Izvestiya: Mathematics*, 63 (4): 79–100.

Mishchenko, E. F. and Rozov, N. Kh., 1980, *Differential Equations with Small Parameters and Relaxation Oscillations*, New York, Plenum Press.

Mishchenko, E. F. and Kolesov, A. Y., 1993, "Asymptotic theory of relaxation oscillations", *Proceedings of the Steklov Institute of Mathematics*, 197: 3–94.

Naidenov, V. I. 1992, "Nonlinear model of the Caspian sea level oscillations", *Mathematical Modeling*, 4: 50–64.

Poincaré , H., 1902, "Sur la stabilité de l'équilibre des figures piriformes", *Philosophical Transactions of the Royal Society of London*, 198, series A.

Pontryagin, L. S., Boltyanskii, V. G., Gamkrelidze, R. V., and Mishchenko, E. F., 1962, The Mathematical Theory of Optimal Processes, New York-London, John Wiley and Sons.

Pontryagin, L. S., 1985, *Selected Works*, Classics of Soviet Mathematics, New York-London-Paris, Gordon and Breach.

Zelikin, M. I. and Borisov, V. F., 1994, *Theory of Chattering Control with Applications to Astronautics, Robotics, Economics, and Engineering*, Boston, Birkhäuser.

Zelikin, M. I. and Zelikina, L. F., 1999, "Exact constants in Kolmogorov-type inequalities", *Proceedings of the Steklov Institute of Mathematics*, 227: 131–139.

Zelikin, M. I., Zelikina, L. F., and Schulze, J., 1987, "On the forecasting of the fluctuation in level of closed lakes", in: *Proceedings of the 1st World Congress of the Bernoulli Society (Prohorov, Yu. and Sazonov, V. V., eds.). Vol. 2, Mathematical Statistic Theory and Applications*, Utrecht, VNU Science Press: 655–658.

8 The Development of Systems Science: Concepts of Knowledge as Seen from Western and Eastern Perpsective

Andrzej P. Wierzbicki

This chapter presents first a historical review of the role of technological concepts and mathematical methods in the development of systems science, recalling the origins of systems analysis and the role of the hard and soft approaches to systems science. This dialectic contradiction is typical for Western science, but also helped to achieve many synthetic approaches and results, particularly proposed of researchers from either Eastern and Central Europe, or from Far East countries. In further sections, the chapter concentrates on the concepts of knowledge and of models on the verge of information society and knowledge-based economy. Knowledge-based economy and information society are defined similarly by knowledge and information becoming an essential or even dominant productive factor. In order to reflect on their impacts, it is essential to understand better the distinction between information and knowledge. Various types of understanding of the concept of knowledge are discussed in this chapter. One type characteristic of hard sciences in information age is related to synthesizing information into mathematical models that can be analyzed by using computers. Data mining in very large data sets is also related to finding patterns or models that synthesize characteristics of data relevant for a given purpose. Most of methodological conclusions related to mathematical modeling and data mining is not restricted to computer science, but have interdisciplinary character and support interdisciplinary research. Knowledge and information become either more commercialized in knowledge-based economy – or, if supported by public funding, more accessible for public use. Thus, it becomes also more important to make knowledge more accessible in the form of mathematical models used by various scientific disciplines. Easy exchange of computerized mathematical models will help in a better verification and validation of research results for knowledge-based economy and information society. However, in order to increase such accessibility, better standards and software tools for analysis of mathematical models should be developed. As an example, this chapter presents the need of such standards in a part of modern systems science – in multi-objective model-based decision support. A survey of reference point methodology and of its applications for model-based decision support systems in engineering and environmental control is included. Conclusions of the chapter relate to the concepts of knowledge and of models from mathematical, technological and humanistic perspectives and to the Western and Eastern traditions of understanding these concepts.

8.1 Historical Perspective:
Hard Versus *Soft* Systems Science

The development of modern systems science, at least from Western perspective, was helped by a dialectic contradiction of "soft" versus "hard" approaches. The origins of systems analysis can be traced back to the works of Auguste Comte (1758–1857) and Herbert Spencer (1820–1903) in the middle of 19th century, when foundations of sociology were created. Both of them postulated a systemic approach to the analysis of social phenomena. This approach was naturally "soft" in this sense that no mathematical models and no hard methodology of analysis were present at that time. Moreover, the approach of Comte contained an essential element of modern soft systems science: it was *holistic*, Comte postulated a comprehensive analysis of all aspects of social life (Abraham 1973).

However, modern systems science developed since the middle of 19th century starting essentially from a "hard" approach or, precisely speaking, from a critique and slight softening of the techniques developed by *operations research*. Operations research was treated first as a management technique and then as a part of management science; it consisted in formulating a mathematical model of a given managerial operation and then optimizing the parameters of this operation (e.g. minimizing its costs by selecting appropriate specific parameters and decisions). *Hard systems analysis* developed from a critical generalisation of this approach. In more complicated, less precisely structured cases, more appropriate problem formulation consists not only in optimizing with respect to parameters, but also in analyzing and selecting the structure of the system, considering many goals, analyzing the nature of the decision process, and so on.

This development was soon countered by the ideas of general systems theory of Ludwig von Bertalanffy (1901–1972), who stressed the soft and holistic approach, cognitive enlightenment or heureka effect, and deliberative decision making (Bertalanffy 1968). This theory developed later considerably as a part of modern sociological approach to systems science.

Such dialectic opposition was treated in a rather paradigmatic sense by Western science: in different disciplines different approaches were pursued, certain journals accepted papers following either this or that paradigm. However, systems science was also taken up by researchers from East and Central Europe just after the fall of Stalinism at the end of the 1950s. Less driven by the imperative "publish or perish", these researchers could depart more widely from paradigmatic schools; many interesting contributions and approaches to systems science resulted from this development. Motivated by different reasons – the Far East tradition of harmonious synthesis – similar synthetic results were developed by Japanese, later also by Chinese and other Far East researchers. We shall comment on these developments in slightly more detail.

Operations Research and Optimization Techniques

For all the development of soft systems thinking, most of applied systems research was based on the hard approach – mathematical modeling and structured decision processes. Mathematical optimization techniques played an important role in this development. In the 1940s, George B. Dantzig (b. 1914) introduced the concept of *linear programming* (formulating and solving optimization problems of linear functions with linear constraints) and the basic technique of solving such problems computationally, the simplex algorithm. In fact, similar problems were considered earlier – e.g. in economic planning by Leonid V. Kantorovich (1912–1986) in USSR – but the results of Dantzig had a turning-point character. They were delivered just before the development of digital computers and motivated many of early applications of computers[1]; they motivated the emergence of *operations research* and *management science*, since the first applications of linear programming by Dantzig concerned essentially a technical, logistic problem (supplying US aircraft carrier fleet on Pacific at the end of Second World War, see chapter 4); and, last but not least, they motivated a tremendous development of the theory and computational techniques of optimization in the last fifty years.

Operations research concentrated on managerial applications of optimization techniques; thus, it concentrated on linear, combinatorial, integer and mixed integer programming. However, many other applications contributed to the development of *nonlinear programming*. Related to the development of dynamic systems theory was *dynamic optimization*. It is an old subject in mathematics, as expressed in variational calculus; however, computer calculations and the needs of controlling missiles motivated the development of so-called *dynamic programming* by Stuart Dreyfus and Richard Bellman (1920–1984), which is only a specific and not necessarily the best dynamic optimization technique. In response, in USSR the so-called maximum principle of Lev S. Pontryagin (1908–1988) was developed that, beside mathematical elegance, provided for more efficient dynamic optimization algorithms and was generalized to many other cases of dynamic models (see chapter 7)[2]. That type of competition happened several times. For example, the simplex algorithm of linear programming might have exponential dependence of computing effort on the dimension of the problem, which limits the usefulness of this algorithm for very large problems, related e.g. to deciphering communication codes. At the end of the 1970s a result of Leonid Khachian from Moscow was announced proving that linear programming has polynomial computational complexity and thus much more efficient algorithms than the simplex algorithm should exist[3]. This created an increased interest in large scale linear programming; but the nonlinear programming algorithm used by Khachian, although good for theoretical proof, was not competitive computationally with the simplex algorithm; neither were other algorithms proposed soon after Khachian in the USA, such as a centre algorithm of Karmarkar. Thus, many years and novel concepts were needed until efficient nonlinear algorithms for large-scale linear programming – the so-called interior point algorithms – were developed through international collaboration (Polish and Israel researchers contributed significantly to this work).

Optimization techniques developed many branches, such as *parametric optimization, nondifferentiable optimization,* and others. Especially important is *stochastic optimization* (necessary today, for example, for financial engineering), which was developed for many years in the West, starting with the foundations of decision theory under uncertainty by John von Neumann (1903–1957) and Oskar Morgenstern (1902–1977); but it received a new impetus with the results of the Kiev school of stochastical approaches. Equally important is vector or *multi-criteria optimization,* essential for considering many goals in systems analysis, that will be considered in more detail in the final sections of this chapter.

Optimization techniques today are called jointly *mathematical programming* and provide for a wide range of tools for analyzing various mathematical models. However, the paradigmatic Western approach of operations research, consistent also with the approach of classical decision theory, considered optimization not as a tool of analysis but as a goal and the defining feature of rational behavior. The behavior of an individual was considered rational if it was based on maximization of utility function by appropriate decisions. Such paradigmatic approach was criticized by many schools of thought, and it led operations research to a crisis at the end of the 1970s. Slogans such as "the future of operations research is past" (Ackoff 1979) expressed this crisis of treating optimization as a goal. The diagnosis of this crisis was helped by contributions from Central Europe (accustomed to crises) and pointed out to many methodological and institutional drawbacks of operations research. One of them was the "instrumental tradition": the necessity to publish results promoted researchers who devized elegant algorithms for specific cases of a detailed field of optimization; later, they looked around for opportunities to apply these elegant instruments, oversimplifying the models of real problems if that was needed for the application. Naturally, a quite opposite approach is methodologically correct: one should construct a relevant mathematical model of a given applied problem and then select a tool – say, an optimization algorithm – from a broader variety of tools. Another drawback of operations research was not enough attention given to the methodology of model building, analysis and validation.

Systems Science: Decision Theory and the Contribution of Technological Disciplines

The essence of systems science is strongly related to the understanding of decision processes. The question how people make decisions is an old and traditional subject of many disciplines: statistics, economics, psychology, and so on. The basic paradigm of economic decision making is based on game theory and cardinal utility functions, given by von Neumann and Morgenstern (1944), as well as on value theory (Debreu 1959). In this paradigm, it is assumed that a rational producer or consumer maximizes her/his utility or value function. Early in the development of systems science, however, this basic paradigm was questioned and criticized from many disciplinary and methodological perspectives. For example, Herbert Simon (1916–2001) questioned the ability of an ordinary decision maker to optimize her/his decisions, and proposed a model of satisficing[4] decision making (Simon 1957).

The development of the hard approach to systems science was stimulated also by technological concepts and engineering tradition, since the methodology of engineering design or production planning can be seen as a prototype of structured decision processes. However, the impact of technological concepts on systems science was even deeper, most notably related to the concept and the methodology of using *mathematical modeling* in systems analysis. The use of mathematical models for expressing knowledge will be the main subject of this chapter; here we give only some introductory, historical comments.

Mathematical models were used also by mathematical economics and game theory, as mentioned above, as well as by operations research. But other, especially technological disciplines contributed sometimes more significantly in this context. Perhaps the most important impact had the concepts of dynamic models, feedback and stability of nonlinear dynamic systems, leading later to modern theory of chaotic behavior. The last has now a very basic importance in understanding the contemporary world and is treated as a discipline on its own. However, all these concepts originated actually in control science (automatic control engineering, automation, cybernetics, and so on), which is based on the study of dynamic feedback systems[5]. Even now, a good understanding of dynamic systems and feedback phenomena requires an education in control science and engineering. A control engineer learns in bachelor-level studies concepts and techniques taught in mathematical economics first in doctoral-level courses. Without such deeper understanding of the behavior of dynamic systems with feedback, no amount of deliberation and holistic concentration could produce an understanding e.g. of some aspects of chaotic behavior.

The usage and the methodology of mathematical modeling was developed by technological disciplines particularly in Central Europe. For example, one can consider a specific Warsaw school of systems science and mathematical modeling, with many achievements in the basic methodology of systems analysis (Findeisen 1985); in diverse branches of systems science, in the Institute of Systems Research of the Polish Academy of Sciences, led until recently by Roman Kulikowski; in developing the concept of rough sets as an alternative to fuzzy set theory (Pawlak 1991); and in the methodology of mathematical modeling regarding multi-objective decision support and a rational theory of intuition (Wierzbicki 1984, 1986, 1997). This school was interdisciplinary, but based mainly on technological expertise.

The need of interdisciplinary research in systems analysis was recognized early. Already in 1973 a political agreement between USA and USSR resulted in the creation of International Institute of Applied Systems Analysis (IIASA) in Laxenburg, near Vienna. Stronger than its political role as a tool for *détente* between the superpowers was its scientific role as an international and interdisciplinary centre for applied systems analysis. The collaboration of scientists from various disciplines and various cultural perspectives provided for many lasting results, especially in the methodology of systems science and in blending hard and soft approaches. For example, already in 1984 an international symposium organised by IIASA stressed the theme of plural rationality and interactive decision processes (Grauer, Thompson, and Wierzbicki 1985), presenting diverse perspectives of what might be called rational and how to organise a decision process. While Western scientists presented mostly the separate viewpoints of paradigmatic either hard economic or soft sociological rationality, Central European and Far East researchers looked more for a new synthesis of hard and soft, even if motivated mostly by techno- logical perspective. The argument for seeing *diversity* as a resource was advanced. This argument is best explained in evolutionary terms: while genetic diversity has been essential as a resource in biological evolution, cultural diversity is necessary for civilisation evolution. We do not know what crises await us in the future, thus we should be prepared by preserving diverse cultural viewpoints.

An example of a synthesis of soft and hard approaches in systems science by Japanese researchers is the "Shinayakana systems approach", which stresses flex- ibility and using diverse approaches as needed (Sawaragi and Nakamori 1992). Another example, actually of Chinese origin – though developed in the USA – is the theory of *habitual domains* by P. L. Yu (1995). Most notable is the initiative of the Japanese Advanced Institute of Science and Technology, Hokuriku, of creating a School of Knowledge Science with the explicit aim to bridge gaps between soft and hard approaches and disciplines. Shinayakana systems approach contributes to the aims of this School. The remarks presented in further sections concern- ing the aims of knowledge science on the verge of knowledge-based economy are based also on the publications of this School. For Western science, such initiatives remain a challenge. For example, some of the ideas of Shinayakana systems approach were included into some aspects of reference point methodology described below, but most of Western science prefers divided schools of thinking.

Soft Approaches to Systems Science and the Role of Intuition in Decision Making

Yet the soft systems theory, as defended by sociologists (we have already stressed that systems theory started as a part of sociology), has its deep merits, best exem- plified by the following quotation "soft models – hard thinking; hard models – soft thinking"[6]. The resolution of the contradiction of soft versus hard systems science is not reached yet, at least in the Western tradition, while the Eastern tradition, which prefers harmony to contradictions, has produced several attempts to har- monise soft and hard systems approaches.

The revival of soft systems thinking started with the attempt of von Bertalanffy to create a general systems theory – that stressed holistic understanding, synergy (the whole is bigger than the sum of its parts), *Gestalt* perception of a problem, the moment of cognitive enlightenment (*eureka* or *aha* effect). This revival has its own history (Jackson 2000) and we shall present only a short outline here.

Soft systems methodology as defined by Peter Checkland (1981) criticises hard systems approach as systems engineering that pre-assumes the existence of a system and starts to improve its functioning along given objectives. In reality, dif- ferent actors might opt for different objectives or even have diverse cultural per- ceptions of what constitutes the problem to be solved. Checkland gives the advice not to start reengineering, but to deliberate on the functioning of the system as a whole – to start *soft systems thinking*. While this is a valid observation, Checkland criticises hard systems analysis as it was formalised by RAND corporation in the early years of system science. Thus he fails to notice that the hard approach, upon encountering even in engineering the difficulties described by him, developed its own methods of dealing with these difficulties. Moreover, his criticism is based on the observation that hard, "engineering" systems approach starts with the assumption of the existence of various systems in analysed reality, while this assumption is actually a presumption. On the other hand, he fails to observe that the concept of a system constitutes a mental model, and people very often confuse models with reality. Indeed, this mistake is more often committed in social sciences than by engineers, who are accustomed to check the realism of their systemic designs in practice. However, when interpreted as a criticism of paradigmatic, analytical decision making, his observations are certainly true: decisions in complex situations are not made according to the economic decision paradigm.

This avenue of criticism is even more convincingly presented in the book *Mind over Machine* (1986) by Hubert Dreyfus and Stuart Dreyfus which shows that the way decisions are made changes with the increasing level of expertise of the decision maker. While a beginner, novice or apprentice needs analysis to support her/his decisions, specialists, experts and master experts either make decisions instantly, intuitively, or rely on deliberation. In the case of operational or repetitive decision making, the empirical evidence for such a thesis is abundant: for example, when driving a car, a novice has to think before shifting gears, while experts make decisions with "entire body". In the case of strategic or creative decision-making, when often new types of decisions are needed, Dreyfuses designed and conducted a special experiment involving playing chess. They saturated the analytical part of the brain of a player with tedious computations in interaction with a computer, while letting her/him at the same time compete at chess. Apprentice players could not play chess when the analytical parts of their minds were saturated, while for chess masters it did not make any difference. The authors express their opinion that these results can be interpreted diversely. They are related to the way we perceive reality by images and *Gestalt*, recognise patterns, to the typical connections of *roads of the brain* established by training, to the functions of *the right hemisphere of the brain* and to the concept of intuition. However, Dreyfuses refuse to analyse the latter concept.

The author of this chapter took up the research at this point: if the way that experts make decisions is deliberative, relies on intuition, then we cannot construct good decision support systems without a rational theory of intuition. Constructing such a theory is not purely soft systems thinking; on the contrary, it is a multidisciplinary task and might start with purely technological arguments. In fact, the decisive argument comes from modern telecommunications and computer science. While the bandwidth needed to transmit images is about 10^2 times larger than that needed to transmit speech, the computational effort needed to process data increases at least with the square of the amount of the data, thus it is approximately 10^4 larger in the case of processing images. When accepting this technological knowledge, we might start a quite different investigation and propose a thought experiment in the sense of Thomas Kuhn (1964) and evolutionary epistemology (Wuketits 1984). We thus ask the questions: how did we, humans, think just before developing language; and what happened to the old, "animal" type of thinking?

Humans, before developing language, were already quite capable of dealing with their environment, hence had quite powerful capacity of processing images and other signals in a holistic way; this was related a subconscious use of multi-valued, fuzzy logic, not of binary logic. The development of language started the accelerated evolution of human civilisation, the intergeneration transmission of knowledge. Binary logic was discovered together with language, in order to convince others about necessary actions (by arguments such as: "we must take this action, otherwise very bad things would happen"), but is not naturally related to the description of a real world. This development put the original, holistic processing of visual and other information further into subconscious parts of our mind; *we still have the capacity of such reasoning, but we call it intuition.*

From these premises, a rational theory of intuition can be developed – rational in the Popperian sense, that is, falsifiable, together with practical falsification tests of theoretical conclusions (Wierzbicki 1997, 2000). We just repeat the conclusion that intuition is subconscious or quasi-conscious processing of the totality of information (hence holistic) when using a part of our brain (probably located in the right hemisphere) specialised in the old, non-verbal, "animal" type of thinking. This explains not only the role of deliberation and soft systems thinking, but produces also constructive advice on how to structure intuitive decision processes, how to stimulate intuitive, cognitive enlightenment, and why Korean archers are rational when using Zen concentration techniques before competing at Olympic games. Note that such rational theory of intuition is neither hard nor soft result, but a multidisciplinary synthesis.

Soft systems theory continues to be developed further, with the emergence of critical systems thinking that aims at reconstituting systems thinking as a unified approach to problem management. This involves, first, showing the complementary role of diverse systems methodologies (where diversity is again recognized as a strength), and, second, demonstrating both theoretical and practical power of systems thinking. M. C. Jackson justly observes that while soft systems thinking is good in theory, practical applications usually need some components of hard systems analysis.

The author of this chapter is deeply convinced that systems approach – necessarily soft and deliberative in the absence of hard models – can be much profitable for understanding the future, particularly today, in times of great and fast changes. Thus, the remainder of this chapter concentrates on more forward-looking analysis of concepts outlined above, particularly those related to knowledge and modeling.

8.2 Information Civilisation: Megatrends and Challenges

The essential aspects of the new economy emerging at the beginning of the new millennium – global information infrastructure, information society and knowledge-based economy – will be considered here jointly as "information civilisation", since their main aspect is information and knowledge becoming a dominant aspect of economic activities. More precisely, information civilisation is understood here as the era of development of information society – or, what is practically equivalent, of knowledge economy (because a common definition of both these concepts stresses the dependence of this society and economy on information and knowledge as the increasingly dominant production factor).

Under the term "megatrends" we understand here important tendencies in social or technical development that can be observed and predicted for a long time horizon. It is important to note that there are reasons to forecast the length of the period of information civilisation for many future decades, perhaps for entire 21st century. The megatrends discussed below will also last for several decades. The reasons for these long-lasting megatrends are as follows. Many new ideas and developments of the contemporary science and technology are not implemented as fast as it would be indicated by the fast progress of science or purely technological reasons. The delays in their implementation result from diverse social and economic reasons. An example of this is the development of digital television. Its theoretical foundations were prepared almost 40 years ago – and a wide implementation is not yet fully achieved. If only technological reasons were decisive, then the time of implementing digital television could be shortened to 20 years; social and economic reasons were responsible for much longer implementation time. There are many other such examples. For these reasons, we can today forecast rather precisely – in qualitative terms – which megatrends will be decisive for the future development of information civilisation; the uncertainty concerns rather the scale, timing of full implementation and future technological details.

The selection of trends that should be recognized as most important and relevant depends on the evaluation of their economic, social, technological and intellectual impacts. From this perspective, three basic megatrends can be selected (Wierzbicki 1999) as decisive for the development of information society and as generators of many resultant trends of more specific character: the technological megatrend of integration or convergence; the social megatrend of changing and creating new professions; and the intellectual megatrend of conceptual challenges.

The megatrend of technological integration or convergence concerns media, methods and systems of transmitting and processing information, but also practical use of information. It starts with a general digitalisation of the methods and systems transmitting and processing information, but includes also the trends of: multimedia communication, mobile telecommunication, fast increase of transmission rates in telecommunication networks, integration of new telematic services into complex service systems, and e-commerce, e-banking, etc. The latter trend illustrates also two facets of this technological megatrend. While the megatrend is based on exploiting the technological advantages and possibilities of integration provided by the digitalisation of all information, it is also driven by market gains and by commercial applications. The same will concern the development of knowledge-based economy: while being enabled by technological integration, it will be driven by gains obtained from commercialisation of knowledge exchange.

Concerning the megatrend of changing and creating new professions, we observe that the development of information civilisation relies on a substitution of old professions by their new versions or entirely new professions. Old professions require much physical effort and are badly equipped in information technology tools; new professions require much knowledge and information and utilise increasingly more information technology tools. This implies an increasing *dematerialization* of work. There are many conclusions related to this megatrend, such as:

a) The formation of new professions and the changes of educational systems necessary for bringing these new professions into effect are the most important social phenomena that limit the speed of development of information society, knowledge-based economy and civilisation.

b) New technologies always give opportunity to get rich – for those, who know how to use this opportunity. However, in the beginnings of the era of information civilisation, this phenomenon is particularly significant: the megatrend of changing professions causes another trend, that of increasing social stratification, sometimes called a *new divide*.

c) The condition of success of information society is human innovativeness in inventing new professions that will provide the employment for majority, not only a small part of population in knowledge economy and will thus limit the impact of the new divide.

d) Many decades of increased demand for education can be predicted. This concerns education on all levels, but in particular university education and life-long continuing education. Together with other technological trends discussed above, this results in a trend of multimedial life-long and continuing education.

Finally, the megatrend of intelectual challenges might be even more important than the previous two: it is formed by the great challenges concerning the way of understanding the world, resulting from the spread of information civilisation. The mechanical way of understanding the world – as a great machine, turning its wheels with the inevitability of celestial matter – will be replaced by a new way, systemic and chaotic. The world will be seen as a great but complex dynamic system, in which there are some laws, but also probable is chaotic behavior, resulting from nonlinear dynamics with strong feedback. A chaotic system is more similar to an avalanche or a tornado than to a big machine; all can happen there, and small changes in initial conditions can essentially change its path.

All what seemed a natural and common sense phenomenon in the old way of understanding the world might be questioned in information civilisation. Thus, we might have many new and great conceptual, intellectual challenges. This concerns quite basic concepts and problems: the way of understanding markets, economy, democracy, human rights, ethical problems, etc. We shall not discuss here this megatrend in more detail. However, it should be noted that the subject of this chapter – especially the role of mathematical modeling – is strongly related to this third megatrend, though it also results clearly from the definition of knowledge-based economy and is related to both megatrends discussed earlier. It should be also noted that precisely the third megatrend constitutes the greatest challenge to our society; without seeing the new society, a knowledge-based economy in new terms, we might miss the most important phenomena or policy aspects.

8.3 Diverse Concepts of Knowledge

When discussing the future of knowledge-based economy, in order to respond to its intellectual challenges, we must understand better the meaning of the term "knowledge". An encyclopaedic definition of knowledge stresses usually at least two aspects of this concept. One understanding of knowledge corresponds to the entire contents of an individual mind, resulting from experience and learning. Another understanding of knowledge stresses its objective and socially utilitarian character: knowledge denotes all such information that results from confrontation with real world, be it incorporated in theoretical reflection or be it not based on a theory, but just useful in applications. In this chapter, we shall consider mostly not the individual, but the social aspect of knowledge.

However, both above definitions are very general, while in the time of information society and knowledge-based economy we need a more technical definition of knowledge. Recall that the known technical definition of the quantity of information contributed essentially to the understanding of this concept, even if various qualitative aspects of information, such as its security, quality, etc., are of more importance today. The concept of knowledge is more demanding that that of information, and we shall consider it from various perspectives, as in the introductory section, soft versus hard, but on broader scale – in so-called soft sciences (humanities and social sciences) versus so-called hard sciences (natural sciences and technology).

Both in soft and hard sciences, we often use the concept of a model. This concept is quite general. Besides its traditional social or professional interpretations, it might mean a small representation or a copy of some object, a pattern on which we base a new version of a system, etc. However, in the professional, scientific language, the concept of a model denotes a synthetic description of a part of reality, constructed for a given purpose. Following the falsification approach in the philosophy of science (Popper 1983), we can interpret any scientific law as a model of reality, valid only subject to failed falsification attempts. Popperian approach to philosophy of science is considered today strict and demanding, while there are also even more relativistic approaches. For example, Paul Feyerabend (1987) denies any fully objective character of knowledge. However, most convincing seems the approach of evolutionary epistemology (Lorentz 1965, Wuketits 1984): objective knowledge and its social transfer are useful for the evolution of humans. We can treat the principle of falsification by Karl Popper as an useful tool for checking the objectivity of knowledge, needed in the evolution of human civilisations. We can thus summarise the above discussions: *models express knowledge while no knowledge is absolute.*

Concepts of Knowledge in the Humanities and the Social Sciences

Humanities have a quite different concept of knowledge from so-called hard sciences. There is actually a duality, an opposition of two different concepts of knowledge in humanities. One of them stresses *model* formation, although in rather general, verbal form: a *ideal type* (of Max Weber) or a *structure* (of Claude Levi-Strauss). The opposite concept stresses not model formation, but hermeneutic *understanding*[7] (with a long history of development of the concepts of hermeneutics, from Edmund Husserl, Hans-Georg Gadamer, Martin Heidegger). This is a very deep dichotomy of basic concepts: knowledge as models versus knowledge as understanding.

On the verge of knowledge-based economy, we need also a deeper though rational explanation of this dichotomy, which can be obtained through the rational theory of intuition outlined earlier. Thus, we need to extend the dichotomy between humanistic models (ideal types, structures) and humanistic understanding (hermeneutics, which can be viewed as intuition formation) to other forms of knowledge and other sciences. We need both the evaluation of various forms of models expressing knowledge and, in order to show limits of knowledge, a rational theory of intuitive reasoning. If we achieve this, we can aim at the development of a *knowledge science*. Such a new discipline should not only help to understand various forms of and limits to knowledge, but also result in a better integration of and transfer of results between diverse disciplines of knowledge and science.

The rational perspective of intuitive reasoning helps also to understand better another distinction between various forms of models expressing knowledge. Hard (nomothetic) sciences and, in particular, engineering, often use mathematical models. However, disciplines such as philosophy, humanities and social sciences – generally called soft (idiographic) – attempt to help in the understanding of an increasingly complex world in various ways. One way is by trying to reach a hermeneutic, intuitive understanding; another way is by using various general types of models - ideal types, structures, other verbal models of difficult issues. Note that even a verbal discussion of such issues, organised in a structured way, is in fact a model that tries to provide certain perspective and to increase understanding. The megatrends of information society, discussed before provide an example of such a model. We might call such verbal models *humanistic models*, as opposed to *mathematical models*.

Soft sciences sometimes attempted – not always necessarily – to increase scientific validity of their statements by trying to incorporate mathematical models from hard sciences. However, humanistic models rely in fact on hermeneutic understanding, on a very high degree of synthesis. From the rational theory of intuition it follows that such degree of synthesis is usually attained by intuitive, holistic perception. Choosing words to formulate such models is difficult, because all words have multiple meanings; on the other hand, we can doubt whether any attempt to make these models harder by trying to express them in a mathematical form is constructive or useful.

An example of such difficulties is the issue of interpreting statistical models of data preferred today by social sciences. These sciences, such as economics, business management or quality management, tend to be fascinated by the possibilities of data processing by computers and by deriving statistical models from these data – without deeper knowledge of the methodology of forecasting that stresses the dangers of interpreting such models. A statistical model is not a causal model, to specify causes and effects in such a model we need deeper, fundamental knowledge, external to the statistical data. The best example illustrating this fact is the strong statistical relation between magnetic storms on Earth and solar spots. Based only on statistical data, one could advance a theory that magnetic storms perturb our vision and thus we see spots on Sun. Such a theory is false, but consistent with the data. In order to specify correctly that solar spots cause magnetic storms, we need a theory of elementary particles emitted by solar eruptions and their impact on the magnetic field of Earth – a deeper theory, external to the data. This example illustrates also a conclusion that it is important to develop knowledge science but it must include the methodology of model formation and interpretation.

Concepts of Knowledge in the Hard Sciences

Hard sciences have their own forms of knowledge: though all use mathematical models, they are by no means united about the definition of knowledge. From the perspective of knowledge engineering – a specific discipline in information sciences – knowledge is defined as a pattern that can be discerned in data. While this definition is very useful for data mining and knowledge discovery in large data sets, it is possibly too narrow for broader applications in knowledge economy. Although knowledge engineering specialists tend to include models as a specific forms of patterns, the interpretation of this relation in other specialities – including computerized decision support – is just opposite: patterns are considered a specific form of models. We shall follow the second interpretation, since the concept of a model is quite general, as commented above.

We already mentioned that in the professional, scientific language, the concept of a model denotes a synthetic description of a part of reality, constructed for a given purpose. The purpose of constructing a model can be diverse, but very often – *e.g.* in model-based decision support – relates to expressing knowledge about a given situation. The forms of models can be also diverse, starting with a general, verbal form and proceeding to more specific forms of mathematical models. In a sense, mathematical models represent today the very basic form of expressing and exchanging knowledge in hard sciences, especially technological knowledge. This is because mathematical models can be easily computerized. We shall discuss this and related concepts in more detail.

8.4 The Importance and Typical Forms of Mathematical Models Expressing Knowledge

A critical element of many scientific investigations – including also such activities as model based decision support – is a mathematical model that represents data and relations that are too complex to be adequately analysed based solely on experience and/or intuition of a researcher or a decision maker.

Models, when properly developed and maintained, can represent not only a part of knowledge of a researcher or a decision maker, but also integrate relevant knowledge available from various disciplines and sources. Moreover, models, if properly analyzed, can help their users to extend their knowledge and intuition. Therefore, the quality of the whole cycle of preparing, maintaining and analyzing models determines to a large extend also the quality of research conclusions or of a practical decision making process for any complex decision problem. There are new developments in modeling methodology and tools that are worth to be discussed in the context of knowledge integration. In addition to the continuously growing opportunities resulting from the progress in database management and in the foundations of modeling, there are many new opportunities which emerge from recent developments in methodologies for and experiences from model-based decision support systems or from the network-based platform-independent software technologies. New opportunities are also offered by new developments in the applications of the World Wide Web for management, policy makers, research and education. All such opportunities are not yet efficiently exploited. There is a need to use all these opportunities to improve the low productivity of model-based work by unifying various representations of a model, facilitating the standardised interfaces to diverse solvers that support different paradigms of model analysis. It is also necessary to improve the possibilities of sharing knowledge represented by various models that can be analysed on heterogeneous hardware available in distant locations.

In order to improve such possibilities, we shall first discuss typical forms of mathematical models. They can be divided into *binary models* (based mostly on binary logical relations, such as patterns discovered in data) and *analytical models*, although this distinction is not always sharp (each model includes logical relations and might include binary ones). However, this general distinction is related in a sense to the basic megatrend of technical integration in information civilisation. This basic trend is based on digitalisation of all forms of information; thus, it is not unnatural to think that the forms of models expressing computerized knowledge will be also closely related to the most basic, binary form of presenting information.

The Importance and Limitations of Binary Models

There are many examples of this trend to replace all other types of information processing by purely binary processing. Genetic algorithms of optimization represent one of such examples; another is a powerful trend towards finding patterns of knowledge in very large data sets. Thus, there is no doubt that binary forms of computerized models encoding knowledge, in a sense natural for implementations on digital computers, will increase their importance in future. For a computer scientist, who tends to see the world as a giant computer, these forms are the most natural models.

This does not mean, however, that this form of models will become a universal or even dominant standard, because binary models have some essential drawbacks, for all their uniformity. First, binary models are in a sense too sharp for representing real world; multi-valued, fuzzy logic expressions and so-called soft computations are much more adequate, but this means already a departure from binary towards analytical models (on the developments and applications of fuzzy logic, see Zadeh 1978 and Zimmermann 1987).

Secondly, each discipline of science uses its own characteristic collection of analytical models which serves as a kind of global language for specialists of this discipline; it would be disadvantageous for the development of science to replace such languages by a universal, binary model form. We should recall here several examples. One of the disciplines that contributed most significantly to the understanding and classification of analytical models was control science, describing the ways of controlling diverse dynamic processes, leading to the development of most of modern technology such as automated manufacturing, robotics etc. But control science also relied on earlier developments of various concepts and models developed in other disciplines, such as the concept of feedback coming originally from telecommunications, or models describing dynamic processes coming originally from mechanics. Control science included binary models in the form of the theory of finite automata, but is much broader and richer than only this theory.

A third drawback regards the fact that the processing of binary models requires usually times that grow faster than polynomially (*e.g.* exponentially) with the dimensions of these models. This is also true for many analytical models, and classical computer scientists usually argue that the processing power of modern computers grows fast enough for not to be troubled by the non-polynomial computation complexity. However, this argument does not take into account the fact that true specialists in complex computations can always find ways to saturate even the most powerful computers. In fact, the advantage of analytical over binary models relies on the possibility of finding special algorithms – admittedly, usually only approximate and only for limited and specific classes of analytical models – that can process these models much faster than in the universal case (that is always doomed to non-polynomial complexity).

Finally, we should note that human mind works quite differently than modern computers, and binary models are a very poor approximation of its operation. A better approximation of human mind are artificial neural networks, but they are not fully adequate yet; we can only say that processing information in human mind is certainly parallel and distributed, but also more complicated than in contemporary artificial neural networks. This argument is also related to the rational definition of intuition discussed earlier in this chapter.

The Role and Challenges of Analytical Mathematical Modeling

Analytical mathematical modeling started as a generalisation of modeling techniques from several disciplines. Models of operations research were augmented with a broader methodological reflection of control science as well as of other disciplines, especially systems analysis. Today, we know well the classification of model types in mathematical modeling: continuous versus discrete, linear versus nonlinear, deterministic versus stochastic, static versus dynamic, open loop versus feedback, single-objective or scalar optimization versus multi-objective or vector one, as well as diverse techniques of dealing with all these model types. This powerful body of knowledge can be developed further as one of basic elements of knowledge science, provided we can respond to several challenges related with the development of information civilisation and knowledge-based economy. Especially important are the two facets of the megatrend of integration: we need more integration of models used in diverse disciplines and we need clearer principles of using commercially knowledge encoded in models. Several challenges can be listed as related to the facet of integration. The first is the need to integrate, in particular, the results from two disciplines almost separate until now: knowledge engineering and mathematical modeling. This is related to the former discussion of the binary and analytical forms of models.

The second challenge regards the need to integrate various methodological approaches to the analysis of diverse types of mathematical models used in various scientific disciplines. We recall that mathematical modeling typically uses model simulation, scenario and sensitivity analysis, single-objective optimization, and sometimes multi-objective optimization and analysis. These approaches might be complemented, however, by multi-objective inverse simulation and scenario analysis. However, all these methodological approaches are not widely known by specialists in various scientific disciplines. While we discuss these issues in more detail in a further section, we note here the related challenge: to make the approaches of mathematical modeling and systems analysis known and usable for diverse scientific disciplines.

The third challenge relates to the fact that computerized mathematical models have diverse standards not only between various scientific disciplines, but also inside each discipline. This makes knowledge exchange cumbersome and difficult; it also impedes comparison of models devoted to the same subject. There exist initiatives aimed at responding to this challenge, such as the idea of a structured modeling language (Geoffrion 1989). However, not only standards of modeling languages should be further developed, but also methods and tools related to model analysis should be integrated within such languages. Thus, much more work is yet to be done before this challenge will be adequately responded to.

Other challenges are related to the facet of commercialisation of knowledge that requires a better understanding and the development of clear guidelines for utilising model-encoded knowledge in several domains: public, commercial and private. The distinction between knowledge that can be considered private or commercial and knowledge that must remain in public domain is a deeply ethical question, one of basic intellectual challenges of the coming era.

8.5 An Example: Computerized Decision Support Systems

Some of necessary developments can be best described on the example of model-based computerized decision support systems. Any decision maker, before making final selection of a decision, wants to understand in the consequences of her/his possible decisions a best possible way. Especially in more complex situations, decision makers typically need help in finding decisions that best correspond to their preferences. These preferences can hardly be precisely defined in advance, because they often change while a decision maker learns about the decision problem.

There are several types of computerized decision support systems used currently for various purposes (Andriole 1989). Most broadly and commercially applied are *data based* decision support systems that use data mining and OLAP techniques to find regularities in large data bases and construct models representing these regularities. Another type are *rule based* expert and artificial intelligence systems, where existing experience and knowledge are described by logical, binary rules. In engineering design and environmental control, but also in operations research and economics, analytical *model based* decision support systems are broadly applied. The latter type takes advantage of knowledge in a given discipline described not by logical, but by analytical models. For example, automatic control theory has a long tradition of using analytical models of control plants and of controllers, while the experience of using fuzzy controllers shows that it is rather difficult to represent these complicated, analytical models by rule based, binary models.

We shall concentrate in this section on analytical model based decision support systems. Even this category is very broad. Such decision support systems can be applied in various disciplinary fields; some approaches stress the advantages of disciplinary, case-oriented tools. On the other hand, we shall show that most of model based decision support problems have interdisciplinary character and thus need more general tools.

Among more general approaches, we can also distinguish several large classes. One is based mostly on *model simulation*: since the models used, say, in automatic control are complicated, we assume some values of parameters and decision variables and just analyze the results of the simulated behavior of the model. Another is based on *single-objective optimization*: we define a criterion or objective function and analyze solutions which are optimal with respect to this function. In the field of automatic control, a known system of tools for using simulation in decision support is MATLAB, augmented for optimization tasks with its library OPTIX (Optimization Toolbox). On the other hand, economic and operations research applications motivated the development of several *algebraic modeling languages* – such as GAMS, AIMMS, AMPL, LANCELOT – which are, in fact, systems of tools for analytical modeling dedicated to single-objective optimization (while it is often difficult to use them for model simulation).

However, experience in model building and analysis, say, for automatic control (Wierzbicki 1984), shows that single-objective optimization can be only a tool of more advanced model analysis, never a goal. This has been confirmed by applications in many other fields of engineering and environmental sciences: good model based decision support systems should rely not on single-objective, but on *multi-objective* or *vector optimization*. Although it is true that all multi-objective optimization can be reduced to parametric single-objective optimization, systems of tools required for multi-objective optimization are much more sophisticated than for single-objective one and simplistic extensions of the latter to include the former are often inadequate (Steuer 1986).

Even when concentrating on analytical model based decision support systems using multi-objective optimization, we can distinguish many approaches (Sawaragi, Nakayama, and Tanino 1985). For *interactive decision support*, where the decision maker can change her/his preferences during the decision process, two broad classes of approaches are useful. One is *goal programming*, starting with the work of A. Charnes and W. W. Cooper (1977); another is a generalisation of goal programming, called *reference point methodology*. Reference point methodology can be further adapted for the purposes of multi-objective model analysis that includes not only multi-objective optimization, but also inverse and softly constrained model simulation. We present shortly this and other aspects of reference point methodology in next section. In this presentation, hard mathematical concepts and language will be used – in order to show that the synthesis of soft and hard approaches needs not to rely on abandoning mathematical tools.

Reference Point Methodology

Reference point approaches were developed starting with research done at the International Institute for Applied Systems Analysis since 1980 specifically as a tool of environmental model analysis, although these approaches have found numerous other applications since that time (Wierzbicki 1980, Kallio, Lewandoski, and Orchards-Hays 1980). Similar or equivalent approaches were soon developed, such the weighted Chebyshev procedure (Steuer and Cho 1983) or the satisficing trade-off method (Nakayama and Sawaragi 1983). Later, it has been pointed out that reference point methods can be considered as generalized goal programming (Korhonen and Laakso 1985).

The main advantages of goal programming are related to the psychologically appealing idea that we should set a goal in objective space and try to come close to it: coming close to a goal suggests minimizing a distance measure between an attainable objective vector (decision outcome) and the goal vector. The basic disadvantage relates to the fact that this idea is mathematically inconsistent with the concept of *vector-optimality* (or *Pareto-optimality*, or *efficiency*). One of the basic requirements – a generally sufficient condition for efficiency – for a function to produce a vector-optimal outcome (when minimized or maximized) is an appropriate monotonicity of this function. But any distance measure is obviously not monotone when its argument crosses zero. Therefore, distance minimization cannot, without additional assumptions, provide vector-optimal or efficient solutions.

Nevertheless, setting a goal and trying to come close to it is psychologically appealing; the problem is "only" how to provide for efficiency of resulting outcomes. There are two ways to do it: either to limit the goals or to change the sense of coming close to the goals.

Trying to limit the set of goals is the essence of the *displaced ideal* method of M. Zeleny (1974). If we select goals that are sufficiently distant from the set of attainable outcomes (in the displaced ideal area, far to North-East when maximizing all objectives), then we can prove that norm minimization will result only in efficient outcomes. This does not depend on what norm[8] we use or what properties the sets of attainable outcomes have. However, such limitation means precisely loosing the intuitive appeal of the goal programming approach: if we can set only unrealistic goals, the approach looses its basic advantages.

Trying to change the sense of coming close to the goal changes the nature of the goal. Reference points are goals interpreted consistently with basic concepts of vector optimality; the sense of "coming close" to such points is rather special and certainly does not mean distance minimisation. It should rather mean: (1) decision outcomes in some sense uniformly close to the given reference point, if the latter is not attainable – while the precise sense of uniform closeness might be modified by demanding that the resulting decisions and their outcomes remain efficient *i.e.* vector-optimal; (2) decision outcomes precisely equal to the given reference point, if the latter is efficient, vector-optimal – which, somewhat simplifying, means attainable without any surplus; or (3) decision outcomes in some sense uniformly better than the given reference point, if the latter is attainable with some surplus – thus inefficient, not vector-optimal (where the sense of uniform improvement can be again variously interpreted).

The first two cases coincide (almost) with goal programming; the third case is, however, essentially different: it means not "coming close" in any traditional sense, but "coming close or better".

This change of the sense of "coming close" is in fact deeply related to the discussion how people make decisions in reality and how computers should support decisions. In turn, this is related to the concept of "satisficing decisions" of Simon (1957). This concept was used to describe how people make actual decisions (particularly in large organisations); the concept of *quasi-satisficing* decisions was used to describe how a computerized decision support system should help a human decision maker (Wierzbicki 1983).

Satisficing decision making can be in fact mathematically represented by goal programming. Upon reaching attainable *aspiration levels* (goals, reference levels), the decision maker ceases to optimize and learns to increase these levels, but usually not for current, only for future decisions. One can ask why; the most probable answer is that decision making processes are difficult and this assumption reflects some inherent human laziness or caution. Many further studies have shown that such a satisficing behavior of a decision maker, though might seem peculiar, is very often observed in practice. In particular, the use of various reference levels by decision makers – such as aspiration levels, but including also *reservation levels*, very important *e.g.* in the theory of negotiations – has been repeatedly confirmed in practice.

Independently, however, from the issue whether a real, human decision maker would (or could, or should) optimize in all cases, we can require a different behavior from a computer. A good computer program supporting decisions through model analysis might optimize, hence should behave like a hypothetical, perfectly rational decision maker. However, there is one important reservation: the program should not "out-guess" its user, the real decision maker, by trying to construct a model of her/his preferences or utility function, but should instead accept simple instructions which characterize such preferences.

Thus, the methodology of reference point approaches assumes that the instructions from a user to the computerized decision support system (DSS) have the convenient form of reference points, including aspiration levels and, possibly, reservation levels – and that the user is not asked how she/he determines the reference points. An essential departure from Simon assumptions and from goal programming techniques, however, is as follows: the methodology of reference point approaches assumes that *the computerized DSS tries to improve a given reference point, if this point is attainable.*

Therefore, the behavior of the DSS – not necessarily that of its user – is in a sense similar to perfect rationality. It does not minimize a norm, but optimizes a special function, called achievement function, which is a kind of a proxy utility or value function (of the DSS) such that the decisions proposed by the DSS satisfy the three cases of "coming close or better" described above. Because of the difference – in the last case of "coming better" – to the satisficing behavior, we call such behavior quasi-satisficing. It can be compared to the behavior of a perfect staff (one staff member or a team of them) that supports a manager or boss, who gives instructions to this staff in the form of reference levels. The staff works out detailed decisions which are guided by the given reference point.

However, being perfect, the staff does not correct attainability estimates (a real, human staff might behave otherwise) and does not report to the boss that the reference point is attainable when it really is not. Instead, the staff proposes decisions that result in outcomes as close as possible to the desired reference point and reports these decisions together with their not quite satisfactory outcomes to the boss. If the reference point is attainable without any surplus, the perfect staff just works out the decisions how to reach this point and does not argue with the boss that a different point and different decisions might be better (if not specifically asked about such opinion). If the reference point is attainable with surplus, the perfect staff does not stop working and start gossiping over drinks (as it would be suggested by Simon's model of satisficing behavior). The perfect staff should rather work out decisions that would result in a uniform improvement of outcomes as compared to reference levels, and propose such decisions together with improved outcomes to the boss. Obviously, only a computer program could behave all times in this perfect, quasi-satisficing manner.

In order to discuss the above general ideas and properties in more detail we need some notation and concepts. We already mentioned that we distinguish two parts of a model of a decision situation. One part, called here a *preferential model*, represents the preferences of the decision maker or DSS user (most often, the real users of decision support systems are not the final decision makers, but their advisors – analysts, modelers, designers etc.). In reference point methodology, the attention is not concentrated on the precise form of a preferential model; on the contrary, it is assumed that the preferential model might change during the decision process and the decision support tools should be flexible enough to accommodate such changes. Therefore, we assume that the preferential model is very general, similar to the partial order of Pareto type (which corresponds just to the desire to maximise all decision outcomes). Only the specifics of this model (say, the selection of decision outcomes to be maximized) might also change during the decision process.

The second part of a model of decision situation is called here a *substantive model* that represents the available knowledge about possible decisions and their possible outcomes. Therefore, we assume here that the general form of a substantive model is:

[1] $y = f(x),\ x \in X_0,$

where $x \in \mathbb{R}^n$ denotes a vector of *decision variables*, X_0 is a *set of admissible decisions* which is usually defined by a set of additional inequalities or equations called *constraints*, $y \in \mathbb{R}^m$ is a vector of *model outputs* or *decision outcomes* which includes also various intermediary variables that are useful when formulating the model, even when determining the constraints – thus, the set X_0 is often defined implicitly.

$Y_0 = f(X_0)$ is called the *set of attainable outcomes*. It should be stressed that this set is not given explicitly (even in the simple case when f is given explicitly) and we can only compute its elements by assuming some $x \in X_0$ and then determining the corresponding $y = f(x) \in Y_0$ by simulating the model. The modeler, when analyzing the substantive model, might specify several model outputs as especially interesting – we call them *objectives* or *criteria* and shall denote by $q_i = y_j$, forming an *objective vector* $q \in \mathbb{R}^k$ – a vector in the *objective space*. While this vector and space might change during the decision process, we shall denote the relation between decisions and objectives by $q = F(x)$; the set $Q_0 = F(X_0)$ is called the *set of attainable objectives*.

Since we can change minimization to maximisation by changing the sign of an objective, we can as well assume that all objectives are, say, maximized. Recall that a *Pareto-optimal* decision and its outcomes are such that there are no other admissible decisions that would improve any objective without deteriorating other objectives. A closely related, but slightly broader and weaker concept is that of *weakly Pareto-optimal* decision and outcomes: these are such that there are no other admissible decisions that would result in a joint improvement of all objectives. This concept is actually too weak for applications, since in practice we do prefer outcomes with only one objective significantly improved.

In fact, even the concept of Pareto-optimality is sometimes too weak for applications, in cases where we could improve significantly one objective component at the cost of an infinitesimally small deterioration of another objective. The (limits of) ratios of improvements and deterioration of objectives, determined at a Pareto-optimal decision, are called *trade-off coefficients*; we define *properly Pareto-optimal* decisions and outcomes as such that the corresponding trade-off coefficients are bounded. Even this concept is too weak for applications, since the mathematical sense of "bounded" means "anything smaller than infinity". Truly important for applications are rather decisions and outcomes which are *properly Pareto-optimal with a prior bound*, i.e. such that a finite bound on trade-off coefficients is *a priori* given and satisfied.

Without describing here more precisely the mathematical details of reference point methodology, we outline here only its most basic features (Wierzbicki 1986). We assume here that each objective can be maximized, minimized or stabilized, that is kept close to a given target. For each objective, *reference levels* in the form of *aspiration levels* $q_a \in \mathbb{R}^k$ (that would be good to achieve) are specified by the modeler. Additionally, *reservation levels* $q_r \in \mathbb{R}^k$ (which should be achieved if it is at all possible) might be also specified. These reference levels will be used as main interaction parameters by which the user of a DSS controls the selection of decisions and their outcomes. The values of these reference levels are subject to reasonability constraints only: given lower q_{ilo} and upper q_{iup} bounds for each objective q_i, we require that $q_{ai}, q_{ri} \in [q_{ilo}; q_{iup}]$ and that $q_{ai} > q_{ri}$ for maximized objectives and $q_{ai} < q_{ri}$ for minimized ones. For so-called *stabilized* outcomes (that should be kept close to given targets, as in goal programming) we can use two pairs of reservation and aspiration levels: one "lower" pair as for maximized outcomes and one "upper" pair as for minimized ones.

A way of aggregating the objectives into so-called *order-consistent achievement function* consists in specifying *partial achievement functions* $\sigma_i(q_i, q_{ai})$ or $\sigma_i(q_i, q_{ai}, q_{ri})$ which should: (a) be strictly monotone consistently with the specified partial order – increasing for maximized objectives, decreasing for minimized ones, increasing below (lower) aspiration level and decreasing above (upper) aspiration level for stabilized ones; and (b) assume value 0 if $q_i = q_{ai}, \forall\, i = 1, ..., k$, if aspiration levels are used alone – or assume value 0 if $q_i = q_{ri} \,\forall\, i = 1, ..., k$ and assume value 1 if $q_i = q_{ai}, \forall\, i = 1, ..., k$, if both aspiration and reservation levels are used. If the aspiration levels are used alone, we just check with the help of the sign of an achievement function, whether they could be reached. In such a case, it is useful to define partial achievement functions with a slope that is larger if the aspiration levels are closer to their extreme levels: thus, for maximized objectives $\sigma_i(q_i, q_{ai})$ $= (q_i - q_{ai})/(q_{iup} - q_{ai})$, and similarly with changed signs and q_{ilo} replacing q_{iup} for minimized and combining both formulae for stabilized ones.

An alternative way is to use piece-wise linear functions, e.g. to change the slope of the partial achievement function depending on whether the current point is above or below the aspiration point, or even if it is below the reservation point (Granat et al. 1996). If the values of $\sigma_i(q_i, q_{ai})$ would be restricted to the interval [0;1], then they could be interpreted as *fuzzy membership functions* $\mu_i(q_i, q_{ai})$ (Zadeh 1978, Seo and Sakawa 1988, Zimmermann and Sebastian 1994) that express the degree of satisfaction of the modeler with the value of the objective q_i.

A partial achievement function can be looked upon as simply a nonlinear transformation of the objective range satisfying some monotonicity requirements. The essential issue is how to aggregate these functions as to obtain a scalarizing achievement function with good properties for vector optimization or multi-objective model analysis. One way of such aggregation is to use fuzzy logic and select an appropriate representation of the "fuzzy and" operator. The simplest operator of this type is the minimum operator, which, however, would result only in weakly Pareto-optimal or weakly efficient outcomes when used for multi-objective analysis. To secure obtaining properly efficient outcomes, we have to augment this operator by some linear part. Moreover, membership functions are not strictly monotone if they are equal to 0 or 1. Therefore, inside a vector optimization system, a slightly different overall achievement function must be used, with values not restricted to the interval [0;1]:

[2]
$$\sigma(q, q_a) = \min_{1 \le i \le k} \sigma_i(q_i, q_{ai}) + \varepsilon \sum_{i=1}^{k} \sigma_i(q_i, q_{ai})$$

In the above equation, $\varepsilon > 0$ is a small positive coefficient that defines a prior bound $M = 1 + (1/\varepsilon)$ on corresponding trade-off coefficients. Actually, this bound limits here trade-off coefficients not between various objectives q_i and q_j, but between their transformed values $\sigma_i(q_i, q_{ai})$ and $\sigma_j(q_j, q_{aj})$; in order to obtain bounds on original trade-off coefficients between q_i and q_j, it is necessary to take into account the current slopes of partial achievement functions. Function $\sigma(q, q_a)$ is called an *order-consistent achievement function*, because it preserves monotonicity and its level sets approximate the partial order.

Simpler versions of order-consistent achievement functions were used originally (Lewandowski and Wierzbicki 1989). Some of such versions can be looked upon as a simplification of function $\sigma(q, u)$. For example, suppose only aspiration levels u_i are used, all objectives are maximized and dimension-free and the partial achievement functions have a simple form $\sigma_i(q_i, q_{ai}) = q_i - q_{ai}$. Then the order-consistent achievement function takes on the form:

[3]
$$\sigma(q, q_a) = \min_{1 \le i \le k}(q_i - q_{ai}) + \varepsilon \sum_{i=1}^{k}(q_i - q_{ai})$$

This function can be seen as a prototype order-consistent achievement scalarizing function. General properties of such functions result in the following characterisation of properly efficient solutions (with a prior bound on trade-off coefficients): (sufficient condition for proper efficiency) for any u (with reasonability constraints) a maximal point q^* of $\sigma(q, q_a)$ with respect to $q \in Q_0 = F(X_0)$ is a properly efficient objective vector with a prior bound on trade-off coefficients and, equivalently, a maximal point x^* of $\sigma(f(x), q_a)$ with respect to $x \in X_0$ is a properly efficient decision with a prior bound; (necessary condition for ε-proper efficiency) for any properly efficient $q^* = F(x^*)$ with appropriate prior bounds on trade-off coefficients, there exists an aspiration point q_a^* such that q^* maximizes $\sigma(q, q_a^*)$ with respect to $q \in Q_0 = F(X_0)$ (or, equivalently, x^* maximizes $\sigma(f(x), q_a^*)$ with respect to $x \in X_0$).

Actually, we can prove even more (Wierzbicki 1996): the decision maker or the user of a decision support system can influence the selection of $q^* = F(x^*)$ Lipschitz-continuously by u (and/or w), except in cases when the set of properly efficient objectives is disconnected. We say that this selection is *continuously controllable*. Moreover, the scaling of the partial achievement functions and the scalarizing achievement function is such that the user can draw easily:

The conclusions on the attainability of aspiration points are the following. If the maximal value of $\sigma(q,u)$ with respect to $q \in Q_0 = F(X_0)$ is below 0, it indicates that the aspiration point is not attainable, $q_a \notin Q_0$ and also that there are no points $q \in Q_0$ dominating q_a, that is $\{q \in \mathbb{R}^k : q \ge q_a\} \cap Q_0 = \emptyset$. If this maximal value is 0 and $q^* = q_a$, it indicates that the aspiration point is attainable and properly efficient. If this maximal value is above 0, then the reservation point is either attainable with surplus or dominated by attainable points. Similar conclusions apply to reservation point, if both reservation and aspiration levels are used.

This property justifies the name "achievement function" since its values measure the achievement as compared to aspiration (and reservation) points. We should also stress that all above considerations and properties are not limited to models with convex sets Q_0. In fact, one of the main advantages of reference point methodology is its applicability to arbitrary, nonconvex models. In particular, this methodology is an excellent tool for multi-objective optimization of discrete and mixed-integer models (Ogryczak, Studziński, and Zorychta 1989).

The achievement function $\sigma(q, q_a)$ is nondifferentiable. Moreover, the maximum of this achievement function is in most cases attained at its "corner", *i.e.* at the point of nondifferentiability. In the case of linear or mixed integer models, the nondifferentiability of the achievement function $\sigma(q, q_a)$ does not matter, since the function is concave and its maximisation can be equivalently expressed as a linear programming problem by introducing dummy variables.

In the case of nonlinear models, however, optimization algorithms for smooth functions are more robust (work more reliably without the necessity of adjusting their specific parameters to obtain results) than algorithms for nonsmooth functions. Therefore, there are two approaches to the maximisation of such achievement functions. One is to introduce additional constraints and dummy variables as for linear models. Another is a useful modification of the achievement function by its smooth approximation, which can be defined *e.g.* when using an l_p norm. We quote here such an approximation for the case of using aspiration point q_a alone, assuming that partial achievement functions $\sigma_i(q_i, q_{ai}) \leq 1$:

$$[4] \qquad \sigma(q, q_a) = 1 - \left(\frac{1}{k} \left(1 - \sum_{i=1}^{k} \sigma_i(q_i, q_{ai}) \right) \right)^{1/p}.$$

We stress again that $\sigma(q, q_a)$, even in its above form, is not a norm nor a distance function between q and q_a; it might be equivalent to such a distance function only if all objectives are stabilized.

Until now we discussed reference points as if they were simple collections of their components. However, for more complicated models – for example, with dynamic structure – it is often advantageous to use *reference profiles* or *reference trajectories* of the same outcome variable changing *e.g.* over time. Suppose that a model describes ecological quality of forests in a region or country, expected demand for wood, forestry strategies and projected prices for some longer time – say, next fifty years because of the slow dynamic of forest growth. The user would then interpret all model variables and outcomes rather as their profiles over time or trajectories than as separate numbers in given years. Mathematically, we can represent such a profile as a vector in a space – say, fifty-dimensional – hence the methodology presented above is fully applicable. From the user point of view, however, it is much easier to interpret model outcomes and their reference points as entire profiles or trajectories.

Multi-objective Modeling

Reference point methods can be used for a wide variety of substantive model types. However, methods of optimization of an achievement function attached to a complicated model depend very much on the model type. Moreover, this concerns even model building: constructing a complicated model is an art and requires a good knowledge not only of the disciplinary field concerned, but also of the properties of models of the particular class. There exist today special software tools for building analytical models, called generally modeling systems or algebraic modeling languages – such as GAMS, AIMMS, AMPL (Brooke, Kendrick, and Meeraus 1988, Bisschop and Entriken 1993, Fourer, Gay, and Kernighan 1993). However, they usually represent the perspective of single-objective optimization and can be adapted to multi-objective model analysis only through additional tricks.

Linear models provide a good starting point in modeling. In the case of large-scale models, a practical way to develop a model is to prepare first a linear version and then augment it by necessary nonlinear parts. In a textbook, the standard form of a multi-objective linear programming problem is usually presented as:

[5]
$$\operatorname*{maximize}_{x \in X_0} \quad (q = Cx \in \mathbb{R}^k)$$

$$X_0 = \{x \in \mathbb{R}^n : Ax = b \in \mathbb{R}^m, 1 \le x \le u\}$$

where "maximize" might either mean single-objective optimization if q is a scalar, or be understood in the Pareto sense, or in some more general multi-objective sense. Note that the standard form above uses the equality form of constraints $Ax = b$ in order to define X_0. Other forms of linear constraints can be converted theoretically to equality form by introducing dummy variables as additional components of the vector x. In the practice of linear programming it is known, however, that the standard form is rather unfriendly to the modeler. Thus, specific formats of writing linear models have been proposed, such as MPS or LP-DIT format (Makowski 1994).

Without going into details of such formats, we shall note that they correspond to writing the set X_0 in the form:

[6]
$$X_0 = \{x \in \mathbb{R}^n : b \le y = Ax + Wy \le b + r \in \mathbb{R}^m, 1 \le x \le u\}$$

where the vector x denotes rather actual decisions than dummy variables, thus x, m, n denote different entities than those implied by the standard textbook form. The model output y is composed of various intermediary variables (hence it depends implicitly on itself, although usually through a lower-triangular matrix W). Essential for the modeler is the freedom to choose any of outputs y_i, including actually decisions x_j, as an objective variable q_i and to use many objectives – not only one as it is typical for algebraic modeling languages.

Even more complicated formats of linear models are necessary if we allow for the repetition of some basic model blocks indexed by additional indices, as in the case of linear dynamic models:

$$X_0 = \{x \in \mathbb{R}^n : w_{t+1} = A_t w_t + B_t x_t; b_t \leq y_t = C_t w_t + D_t x_t \leq b_t + r_t \in \mathbb{R}^m,$$
[7]
$$l_t \leq x_t \leq u_t; t = 1,\ldots,T\}$$

where w_t is called the *dynamic state* of the model (the initial condition w_1 must be given), the index t has usually the interpretation of (discrete) time, and $x = (x_1, \ldots, x_T)$ is a *decision trajectory* (called also *control trajectory*). Similarly, $w = (w_1, \ldots, w_{T+1})$ is a *state trajectory* while $y = (y_1, \ldots, y_T)$ is the *output trajectory*. Actually, the variable w should be considered a part of the vector y (it is an intermediary variable, always accessible to the modeler). It is denoted separately because of its special importance – e.g. when differentiating the model, we must account for the state variables in a special way (Wierzbicki 1984). Other similarly complicated forms of linear and dynamic models result e.g. from stochastic optimization. Multi-objective optimization of dynamic models has been also studied extensively (Wierzbicki 1988b).

A modeler who has developed or modified a complicated (say, dynamic) large scale linear model should first validate it by simple simulation – that is, assume some common sense decisions and check whether the outputs of the model make also sense to her/him. Because of multiplicity of constraints in large-scale models it might, however, happen that the common sense decisions are not admissible (in the model); thus, even simple simulation of large-scale linear models might be actually difficult (and is usually not supported by algebraic modeling languages).

An important help for the modeler can be *inverse simulation*, in which she/he assumes some desired model outcomes $y_a \in \mathbb{R}^m$ and checks – as in the classical goal programming – whether there exist admissible decisions that result in these outcomes. *Generalized inverse simulation* consists in specifying also some reference decision $x_a \in \mathbb{R}^n$ and in testing whether this reference decision could result in the desired outcomes y_a. This can be written in the goal programming format of norm minimisation, while it is useful to apply the augmented Chebyshev norm. We will, however, changed signs below, because we keep to the convention that that achievement functions are usually maximized while norms are minimized:

[8]
$$\sigma(q, q_a) = -(1-\rho)\left(\max_{1 \leq i \leq k} |x_i - x_{ai}| + \varepsilon \sum_{i=1}^{k} |x_i - x_{ai}| \right)$$
$$-\rho\left(\max_{1 \leq i \leq k} |y_i - y_{ai}| + \varepsilon \sum_{i=1}^{k} |y_i - y_{ai}| \right)$$

The coefficient ρ indicates the weight given to achieving the desired output versus keeping close to reference decision. It is assumed for simplicity sake that all variables are already re-scaled to be dimension-free.

A vector optimization system based on reference point methodology can clearly help in such inverse simulation. In such a case, we stabilise all outcomes and decisions of interest and use for them partial achievement functions of the form $\sigma_i(y_i, y_{ai})$ or $\sigma_i(x_i, x_{ai})$ similar to those defined earlier in terms of objectives q_i.

An overall achievement function has then the form:

[9]
$$\sigma(q, q_a) = -(1 - \rho)\left(\min_{1 \leq i \leq k} \sigma_i(x_i, x_{ai}) + \varepsilon \sum_{i=1}^{k} \sigma_i(x_i, x_{ai}) \right)$$

$$-\rho\left(\min_{1 \leq i \leq k} \sigma_i(y_i, y_{ai}) + \varepsilon \sum_{i=1}^{k} \sigma_i(y_i, y_{ai}) \right)$$

It is more convenient for the modeler, if such functions are defined inside the decision support system which also has a special function *generalized inverse simulation*, prompting her/him to define which (if not all) decisions and model outputs should be stabilized and at which reference levels.

Even more important for the modeler might be another interpretation of the above function, called *simulation with elastic constraints* or *softly constrained simulation*. Common sense decisions might appear inadmissible for the model, because it interprets all constraints as *hard* mathematical inequalities or equations. On the other hand, it should be stressed that it is a good modeling practice to distinguish between *hard constraints* that can never be violated and *soft constraints* that in fact represent some desired relations and are better represented as additional objectives with given aspiration levels. Thus, in order to check actual admissibility of some common-sense decision x_a, the modeler should answer first the question that constraints in her/his model are actually hard and which might be softened and included in the objective vector. Thereafter, simulation with elastic constraints might be performed by maximizing an overall achievement function similar as above.

Even less developed than user-friendly standards of defining linear models are such standards for nonlinear models. While there exist some standards for specific nonlinear optimization systems – such as in MINOS, GAMS, AIMMS, AMPL – they are devised more for single-objective optimization purposes than for multi-objective modeling and analysis. On the other hand, experience in modeling shows that a model should be analysed multi-objectively even if it is later used for single-objective optimization only. A useful standard was developed in the multi-objective nonlinear optimization system DIDAS-N (Kręglewski et al. 1988, Granat et al. 1994). However, even for nonlinear or nonlinear dynamic models, softly constrained and generalized inverse simulation can be applied similarly as in the linear case.

Applications of Reference Point Methods

The reasoning presented in previous sections might seem rather abstract. Nonetheless, the development of reference point methods was very much applications-oriented. It started with original work on forestry models (Kallio, Lewandoski, and Orchards-Hays 1980) and includes many other applications to energy, land use and environmental models at IIASA. Other applications include satisficing trade-off methods (Nakayama 1983) used for engineering design, and various applications of *Pareto Race* (Korhonen and Laakso 1985). Recent applications of reference point method have been developed at IIASA using a modular tool MCMA (MultiCriteria Model Analysis, Granat and Makowski 1998). These applications relate to regional management of water quality (Makowski, Somlyódy, and Watkins 1996), land use planning (Antoine, Fischer, and Makowski 1997), and urban land-use planning (Matsuhashi 1997). For other applications of multi-objective optimization to environmental problems see also Janssen (1992).

In all these applications to environmental problems, the decisive issue when developing appropriate models was a combination of economic and physical or biological knowledge. For example, in regional management of water quality, the issue is to choose waste treatment technologies that are not too expensive – both in investment costs and in operating costs – but result in only slight violations of water quality standards in an entire river basin. These standards are treated as soft constraints or additional objectives; if replaced by hard constraints, with the requirement of a precise satisfaction of water quality standards, they result usually in much more expensive solutions. For such interdisciplinary modeling, it is often impossible to determine a single objective function that would express all concerns of a modeler; and for a multi-objective formulation, the reference point approach gives a flexible tool of model analysis.

Similar conclusions can be drawn from engineering applications. Here we describe shortly two examples: one application to engineering design and another to ship navigation support.

The first case concerns a classical problem in mechanical design – the design of a spur gear transmission unit (Osyczka 1994). The design problem consists in choosing some mechanical dimensions (the width of the rim of toothed wheel, the diameters of the input and output shafts, the number of teeth of the pinion wheel, etc.) in order to obtain a *best design*. However, there is no single measure of the quality of design of such a gear transmission. We try to make the unit as compact as possible. This can be expressed by minimizing the volume of the unit while satisfying various constraints related to mechanical stresses and to an expected lifetime of efficient work of the gear unit. But we should take into account other objectives, such as the distance between the axes or even the width of the rim of toothed wheel (which might be, at the same time, a decision variable). However, some of these objectives are in fact proxies for economic objectives, such as material costs or production costs.

Another application example shows the usefulness of including dynamic formats of models. This case concerns ship navigation support (Smierzchalski and Lisowski 1994): the problem is to control the course of a ship in such a way as to maximise the minimal distance from possible collision objects while minimizing the deviations from the initial course of the ship. This is a dynamic problem, with the equations of the model described initially by a set of nonlinear differential equations for $t \in [0;T]$. In these equations, the decision or control variable is the course of "our" ship, while the positions and courses of other ships must be also modelled, with initial values of ship positions given as the state vector w_1. Between other model outcomes, the objectives can be modelled as q_1 that represents the (squared) minimal distance between ships and other collision objects, that should be maximized, and q_2 represents the (squared) average deviation from initial course, that should be minimized. To be used in a typical modeling system (the DIDAS-N system was actually used), this model has to be discretised in time. This example shows the practical sense of using dynamic models with multi-objective analysis and optimization.

All these examples confirm the original intention that reference point optimization methods constitute not a specific hard approach in systems analysis, but provide a variety of hard tools to support possibly softest analysis of substantive models about given decision situation. Reference point methodology does not assume any prescriptive way or process of decision making. If the user want to be supported in reaching the maximum of her/his utility or value function, this is possible also in reference point methods (Wierzbicki 1997). However, typical applications of reference point techniques aim only to help in a comprehensive examination of substantive models in order to enhance the intuition of the decision maker and his ability to reach a decision in a deliberative, soft systems thinking fashion.

Notes

[1] Thus, the word *programming* in *linear programming* motivated the name *computer programming*. Even today, solving linear programming problems contributes to a substantial part of computational load of all non-personal computers.

[2] For example, the author of this text introduced a version of maximum principle for dynamic models with control delays – a case that cannot be easily handled by dynamic programming technique (Wierzbicki 1970).

[3] In fact, Khachian based his proof on an algorithm of nonlinear programming by Naum Shor from Kiev.

[4] This is not a misprint, but a neologism introduced by Simon.

[5] The concept of feedback was initiated in telecommunications and control science also originated as a part of telecommunications (see chapters 5 and 6). Many economists or sociologists do not really know what is control science and tend to ask "control? of what?" Similarly, some sociologists do not perceive that so called *system dynamics* approach is nothing else as an adaptation of a very old tool of control sciences – the block-diagrams of dynamic systems – that was used by control engineers long before Jay Forrester popularized it in economics and social sciences.

[6] Due to Harold Barnett in a discussion at IIASA around 1980.

[7] In fact, the postulates of soft systems thinking described in an earlier section are nothing else as the demand for a hermeneutic understanding.

[8] Precisely speaking, Chebyshev norm might result in weakly efficient outcomes.

References

Abraham, J. H., 1973, *Origins and Growth of Sociology*, London, Harmondsworth.

Ackoff, L. R., 1979, "The future of operations research is past", *Journal of the Operations Research Society*, 30: 93–104

Antoine, J., Fischer, G., and Makowski, M., 1977, "Multiple criteria land use analysis", *Applied Mathematics and Computation*, 83: 195–215.

Andriole, S. J., 1989, *Handbook of Decision Support Systems*, Blue Ridge Summit (Pa.), TAB Professional and Reference Books.

Bertalanffy, L., 1968, *General Systems Theory: Foundations, Development, Applications*, New York, Braziller.

Bisschop J. and Entriken, R., 1993, *AIMMS, The Modeling System*. Haarlem, Paragon Decision Technology.

Brooke A., Kendrick, D., and Meeraus, A., 1988, *GAMS, A User's Guide*, Redwood City, The Scientific Press.

Charnes, A. and Cooper, W. W., 1977, "Goal programming and multiple objective optimization", *Journal of the Operations Research Society*, 1: 39–54.

Checkland, P. B., 1981, *Systems Thinking, Systems Practice*, Chichester, J. Wiley.

Debreu, G., 1959, *Theory of Value*, New York, J. Wiley.

Dreyfus, H. and Dreyfus, S., 1986, *Mind Over Machine: The Role of Human Intuition and Expertise in the Era of Computers*, New York, Free Press.

Feyerabend, P., 1987, *Farewell to Reason*, New York, Verso Publishing Co.

Findeisen, W. (ed.), 1985, *Systems Analysis: Foundations and Methodology* (in Polish), Warsaw, PWN.

Fourer R., Gay, D. M., and Kernighan, B. W., 1993, *AMPL. A Modeling Language for Mathematical Programming*, San Francisco, The Scientific Press.

Geoffrion, A., 1989, "The formal aspects of structured modeling", *Operations Research*, 37 (1): 30–51.

Granat J. and Makowski, M., 1988, *ISAAP - Interactive Specification and Analysis of Aspiration-based Preferences*, Interim Report IR-98-52, Laxenburg, International Institute for Applied Systems Analysis.

Granat, J. and Wierzbicki, A. P., 1996, "Interactive Specification of DSS User Preferences in Terms of Fuzzy Sets", *Archives of Control Sciences*, 5: 185–201.

Granat, J., Kręglewski, T., and Paczyński, J., and A. Stachurski, 1994, *IAC-DIDAS-N++ Modular Modeling and Optimization System. Part I: Theoretical Foundations, Part II: Users Guide*, Internal Report, Warsaw, Institute of Automatic Control, Warsaw University of Technology.

Grauer, M., Thompson, M., and Wierzbicki, A. P. (eds.), 1985, *Plural Rationality and Interactive Decision Processes*, Berlin-Heidelberg, Springer-Verlag.

Jackson, M. C., 2000, *Systems Approaches to Management*, Dordrecht, Kluwer Academic Publishers.

Janssen, R., 1992, *Multi-objective Decision Support for Environmental Management*, Dordrecht, Kluwer Academic Publishers.

Kallio, M., Lewandowski, A., and Orchard-Hays, W., 1980, *An Implementation of the Reference Point Approach for Multi-Objective Optimization*, Working Paper WP-80-35, Laxenburg, International Institute for Applied Systems Analysis.

Korhonen, P. and Laakso, J., 1985, "A Visual Interactive Method for Solving the Multiple Criteria Problem", *European Journal of Operational Research*, 24: 277–287.

Kręglewski, T., Paczyński, J., Granat, J., and Wierzbicki, A. P., 1988, *IAC-DIDAS-N: A Dynamic Interactive Decision Support System for Multicriteria Analysis of Nonlinear Models with a Nonlinear Model Generator Supporting Model Analysis*, Working Paper WP-88-112, Laxenburg, International Institute for Applied Systems Analysis.

Kuhn, T. S., 1981, "A function of thought experiments", in: *Scientific Revolutions* (Hacking, I., ed.), Oxford, Oxford University Press (originally published in *L'aventure de la science, Mélanges Alexandre Koyré*, vol. 2, Paris, Hermann, 1964: 307–334).

Lewandowski, A. and Wierzbicki, A. P. (eds.), 1989, *Aspiration Based Decision Support Systems*, Berlin, Springer.

Lorentz, K., 1965, *Evolution and Modification of Behavior: A Critical Examination of the Concepts of the "Learned" and the "Innate" Elements of Behavior*, Chicago-London, The University of Chicago Press.

Makowski, M., 1994, *LP-DIT, Data Interchange Tool for Linear Programming Problems, version 1.20*, Working Paper WP-94-36, Laxenburg, International Institute for Applied Systems Analysis.

Makowski, M., Somlyódy, L., and Watkins, D., 1996, "Multiple criteria analysis for water quality management in the Nitra basin", *Water Resources Bulletin*, 32: 937–951.

Matsuhashi, K., 1997, *Application of Multi-Criteria Analysis to Urban Land-Use Planning*, Interim Report IR-97-091, Laxenburg, International Institute for Applied Systems Analysis.

Nakayama, H. and Sawaragi, Y., 1983, "Satisficing trade-off method for multi-objective programming", in: *Interactive Decision Analysis* (Grauer, M. and Wierzbicki, A. P. eds.), Berlin, Springer-Verlag: 113–122.

von Neumann, J., and O. Morgenstern, 1944, *Theory of Games and Economic Behavior*, Princeton. Princeton University Press.

Ogryczak, W., Studzinski, K., and Zorychta, K., 1989, "A generalized reference point approach to multi-objective transshipment problem with facility location", in: Lewandowski and Wierzbicki 1989: 230–250.

Osyczka, A., 1994, "C Version of Computer Aided Multicriterion Optimization System (CAMOS)", in: *Workshop on Advances in Methodology and Software in DSS*, Laxenburg, International Institute for Applied Systems Analysis.

Pawlak, Z., 1991, *Rough Sets: Some Aspects of Reasoning About Knowledge*, Dordrecht Kluwer.

Popper, K., 1983, *Realism and the Aim of Science*, London, Hutchinson.

Sawaragi, Y., Nakayama, H., and Tanino, T., 1995, *Theory of Multi-objective Optimization*, New York, Academic Press.

Sawaragi, Y. and Nakamori, Y., 1992, "Shinayakana systems approach in modeling and decision support", in: *Proceedings of the 10th International Conference on Multiple Criteria Decision Making*, vol. I, Taipei: 77–86.

Seo, F., Sakawa, M., 1988, *Multiple Criteria Decision Analysis in Regional Planning*, Dordrecht, Reidel.

Shimemura, E., 1999, *JAIST – Japan Advanced Institute of Science and Technology, Hokuriku 1990. School of Knowledge Science, Faculty Profiles*, Ishikawa, JAIST, Asahidai, Tatsunokuchi.

Simon, H. A., 1957, *Models of Man*, New York, Macmillan.

Śmierzchalski, R. and Lisowski, J., 1994, *Computer simulation of safe and optimal trajectory avoiding collision at sea*, Joint Communication, WSM Gdynia and Hochschule Bremerhaven.

Steuer, R.E. and Choo, E.V., 1983, "An interactive weighted Tchebycheff procedure for multiple objective programming", *Mathematical Programming*, 26: 326–344.

Steuer, R.E., 1986, *Multiple Criteria Optimization: Theory, Computation, and Application*, New York, Wiley.

Toffler, A., 1980, *The Third Wave*, New York, W. Morrow.

Wierzbicki, A. P., 1980, "The use of reference objectives in multi-objective optimization", in: *Multiple Criteria Decision Making; Theory and Applications* (Fandel, G. and Gal, T., eds.), Berlin, Springer-Verlag: 468–486.

Wierzbicki, A. P., 1983, "A mathematical basis for satisficing decision making", *Mathematical Modeling*, 3: 391–405.

Wierzbicki, A. P., 1984, *Models and Sensitivity of Control Systems*, Amsterdam, Elsevier-WNT.

Wierzbicki, A. P., 1986, "On the completeness and constructiveness of parametric characterizations to vector optimization problems", *OR-Spektrum*, 8: 73–87.

Wierzbicki, A. P., 1988, "Dynamic Aspects of Multi-Objective Optimization", in: *Multiobjective Problems of Mathematical Programming* (Lewandowski, A. and Volkovich, V., eds.), Berlin, Springer-Verlag: 154–174.

Wierzbicki, A. P., 1996, "On the Role of Intuition in Decision Making and Some Ways of Multicriteria Aid of Intuition", *Multiple Criteria Decision Making*, 6: 65–78.

Wierzbicki, A. P., 1997, "Convergence of Interactive Procedures of Multi-objective Optimization and Decision Support", in: *Essays in Decision Making* (Karwan, M.H., Spronk, J., and Wallenius, J., eds.), Berlin, Springer-Verlag: 19–47.

Wierzbicki, A. P., 1999, "Megatrends of Information Civilisation", in: *International Conference on Research for Information Society*, Warsaw.

Wierzbicki, A. P., 2000, "A Rational Theory of Intuition in Decision Making and its Relation to Negotiation Techniques", *Conference on Group Decisions and Negotiation*, Glasgow.

Wierzbicki, A. P., Makowski, M., and Wessels, J., 2000, *Model-Based Decision Support Methods with Environmental Applications*, Dordrecht, Kluwer Academic Publishers.

Wierzbicki, A. P., Wydro, K. B., and Zieliński, A., 1997, "The era of information civilisation and the developmental role of telecommunications" (in Polish), Przegląd Telekomunikacyjny i Wiadomości Telekomunikacyjne (Warsaw), 5.

Wuketits, F. M., 1984, "Evolutionary epistemology – a challenge to science and philosophy", in: *Concepts and Approaches in Evolutionary Epistemology* (Wuketits, F. M., ed.), Dordrecht, D. Reidel.

Yu, P. L., 1995, *Habitual Domains: Freeing Yourself from the Limits on Your Life*, Kansas City, Highwater Editions.

Zadeh, L. A., 1978, "Fuzzy sets as a basis for a theory of possibility", *Fuzzy Sets and Systems*, 1: 3–28.

Zeleny, M., 1974, "A concept of compromise solutions and the method of the displaced ideal", *Computational Operations Research*, 1: 479–496.

Zimmermann, H. J., 1987, *Fuzzy Sets, Decision Making, and Expert Systems*, Dordrecht, Kluwer Academic Publishers.

Zimmermann, H. J. and Sebastian, H.-J., 1994, "Fuzzy design – integration of fuzzy theory with knowledge-based system design", in: *Fuzzy-IEEE 94 Proceedings*, Orlando, Pergamon Press: 352–357.

9 Coping With Complexity in the Management of Organized Systems

Mario Lucertini

The evolution of industrialized societies in the 20th century has led to the diffusion of systems for the production of goods and services with a high degree of complexity linked to the large scale and/or the high level of automation and information technology. The idea of scientific management, which emerged at the beginning of the century, clearly indicated the need for an explicit, rational "functional representation" of the new production systems, which would replace the implicit, intuitive operational knowledge typical of the traditional workshop. This kind of representation would be helpful in overcoming the intrinsic uncertainty regarding the behavior of the system, which cannot be explained on the sole basis of the behavior of its single components.

Complexity has been recognized as a fundamental characteristic of modern organizational structures, such as government administrations or military organizations. The idea of developing a set of quantitative methods for the organization and managing of production and operations systems took shape in the years around the Second World War: the functional study of processes as being linked with pre-established objectives and the development of a theory of decision-making was identified as a suitable way of coping with the uncertainty due to complexity.

The aim of this chapter is to discuss some ideas on the way to manage complexity, introduced by managers and organized systems designers. Two basic ways to deal with complex systems are considered: *autonomous agents* and *open system control*. Both are integrated decentralized approaches to the management of different kinds of complexity with a different philosophical background. The chapter investigates the various meanings of the term complexity, the corresponding background and the way organized systems have structured themselves to manage complexity. Some consequences for the decision process and for the whole organization system are briefly outlined.

An *autonomous agents* network is a system formed by relatively independent units, which are individually responsible for their actions, and can exchange information (and possibly materials) and negotiate decisions. An *open system* is an individual entity or a network characterized by strong interactions with an only partially known environment. There are several concepts related to these two organization patterns widely used by the people concerned. Some of them are introduced and discussed in the chapter, accompanied by the attempt to introduce some, hopefully meaningful, industrial examples. The relative *open network* structure is the basic frame for most modern organized systems. The theoretical framework developed to analyze open networks is a cornerstone of modern management.

9.1 Forms of Complexity

In the evolution of management science and control systems, the term complexity has assumed various meanings, derived from different fields related to basic science, economics and engineering. The term complexity in those fields takes on different meanings, which can sometimes be interpreted as different aspects of a general frame; and sometimes as symptoms of emerging new problems.

From the point of view of the basic conceptual approaches to the analysis, design and management of complex systems, there are a few basic general frames, which, as a general rule, correspond to different fields of science and technology.

A first frame comes from physics: here complexity is a synonym of *difficult to forecast*, i.e. it is difficult to describe analytically the overall behavior of a given system. Therefore, complex systems are those that are difficult to simulate. The complexity is generally due to a small number of basic factors: there are a huge number of parts (maybe very simple) with strong relationships among them, there are complex nonlinear relationships among some relevant parameters of the system, there are some portions of the system that are difficult to describe in analytical terms, there are significant parameters of the system that are difficult to measure, and there are intricate relations with the environment that are difficult to quantify.

A second frame comes from engineering and is a synonym of *difficult to control*; i.e. it is difficult to obtain a prescribed behavior from a given system. The system may even be quite simple, a good model may exist and a reliable simulation may be available (therefore good forecasts are available), but the so-called inverse problem is difficult to solve and the usual requirements imposed on the overall behavior are difficult to match. Even when the inverse problem can be solved, it may be difficult to find a decision process and the corresponding information flows that are able to produce the required outputs. The practical difficulty of translating solutions obtained by an optimization process into procedures to be implemented in real contexts often decides the success of many advanced reorganization projects. Decentralized systems and local search techniques generally allow simpler implementation rules, while centralized hierarchical organizations and global optimization produce an overwhelming *a priori* coordination activity that is often of no use in effectively managing the system or in ensuring it performs correctly. On the other hand, there are a vast set of models for centralized management (generally based on complete information, a single decision maker, one or more objectives and complete rationality), whereas a multi-decision-making environment requires a deeper analysis, more sophisticated models (generally based on incomplete information and bounded rationality) and a careful evaluation of information flows, negotiation processes and time lags. The multi-decision-making context is in any case an essential requirement when we include the user or client environment in our analysis. Indeed, while typical in-house analysis includes the organization structure, the operators' behavior, and the technology available (all elements under the control of the company management), a user-oriented analysis introduces elements outside the company control, such as client preferences, market alternatives and price-quality tradeoff.

A third frame comes from mathematics and computer science and is a synonym of *difficult to compute*; i.e. it is difficult to carry out, in a reasonable span of time and within a reasonable amount of memory, all the operations needed to find an answer to a given question. The problem is generally completely defined and an algorithm to find a solution is known, but the number of operations that a computer must execute (*computational complexity*) grows with the size of the problem and became easily too high to be completed within a reasonable amount of time. A well-known example is when the computational complexity of the best-known algorithm grows exponentially with the size of the problem. The theory of NP-completeness, developed in the 1970s from the work of Stephen A. Cook (1971) and Richard M. Karp (1972), soon became a standard in algorithm analysis and spread rapidly over almost all scientific and technical fields (Garey and Johnson 1979).

More recently, the evolution towards more comprehensive models has emphasized the difficulty, well known to people involved in forecasting and simulation, of understanding the behavior of a system, because of the strong dependence of internal behavior on exogenous elements and the lack of a theory allowing a quantitative description of some portion of the system (*difficult to model*). This is a well known problem in biology, in economics and in social systems. In production environments it is typically related to the management of organized systems where the human element and its interactions with the environment play an important role. In this framework, when *bounded rationality* elements enter into the description of the system, also the other simulation, forecasting and control aspects become difficult to formulate correctly in logical terms.

People involved in the design of a production system, in technology innovation and in management, have to cope with a number of complexity problems and have used different methods to simplify the problem and to seek acceptable solutions.

In order to go more deeply into the aspect of the complexity related to modeling and to cope with limited information and bounded rationality, it is useful to introduce a distinction between *process complexity* and *system complexity*. The former refers to the behavior of one or more well identified components, whereas the latter is related to the structure or architecture of a system where many objects, a few of which may not be identified, flow and interact in order to match one or more local and/or global goals. In the case of networks, if the network is given and the analysis concerns for example the flows, we are dealing with a process problem; whereas, if the topology of the network, the interaction protocols and the connections with the environment are the objects of the analysis, we are dealing with a system problem. The latter is typically the case of biology, but also of most network flow systems, where the model describing the behavior is built up by looking to sections of the system defined on a local basis and not on the basis of the overall single paths of the flow elements going through a huge number of parts of the network, often far from each other.

In practice, the boundary between process and system design is somewhat vague. In several production process-design problems, the human elements have a fairly acceptable functional description and many network design problems are formulated in a well-defined demand environment. On the other hand, several flow design problems require complex behavioral models and many scheduling problems must take into account the complex ways in which the people involved operate as well as the intricate environmental constraints.

Many conceptual frames and practical tools have been developed over the last century to deal with process complexity (sometimes with significant attempts also to tackle system complexity aspects). Among others: Frederick W. Taylor (1856–1915) scientific management (Taylor 1911); Henry Ford (1863–1947) assembly line (Ford 1922); Henri Laurence Gantt (1861–1919) scheduling theory (Gantt 1910, 1919); Alfred P. Sloan (1875–1966) multi-divisional organization (Sloan 1963); John von Neumann (1903–1957) sequential machine (Burks, Goldstein, and von Neumann 1946); Taiichi Ohno (1912–1990) just in time (Ohno 1988); linear programming, with the contributions by Wassily W. Leontief (b. 1906), Leonid V. Kantorovich (1912–1986), Tjalling C. Koopmans (1910–1985), Georg B. Dantzig (b. 1914), Paul A. Samuelson (b. 1915), among many others (Koopmans 1951); Lester Randolph Ford (b. 1927) and Delbert Ray Fulkerson network flow theory (Ford and Fulkerson 1962); John L. Burbidge (1915–1985) group technology (Burbidge 1996); and Herbert Simon (1916–2001) analysis of non programmed decisions and bounded rationality (Simon 1982).

As far as system complexity is concerned, in spite of the many (sometimes successful) attempts to develop a system framework from a process oriented approach, only a few attempts have been made to build a general theoretical approach (some of them with the flavor of a new religion). Among others: Taylor scientific management with his disciples and followers; Henri Fayol (1841–1925) general principles of management (Fayol 1918, 1949); Elton Mayo (1880–1949) analysis of human and social problems of an industrial civilization (1945); W. Edwards Deming (1900–1993) total quality approach and 14 points for management with his followers (Deming 1950, 1986, 1993); Kaoru Ishikawa (1915–1989) house of quality (Ishikawa 1976, 1990); and Gen'ichi Taguchi (b. 1924) tools for quality (Taguchi and Konishi 1987).

9.3 The Forms of Simplification

The question considered in this section is basically the following: how have organized systems structured themselves in order to manage (the different kinds of) complexity?

The statement above invites the natural question of when the structure devoted to the management of complexity is endogenous, internally born, and when it is exogenous, the outcome of an external action.

Large organizations, high-tech units, and objective-oriented high-performing groups tend to build up internal processes to manage complexity, such as decomposition and standardization processes, thereby developing specific professional skills. On the other hand, the reference environment for people concerned with complexity and the structures dedicated to the management of complexity, are seldom totally internal, but different, more or less independent, professional organizations or others external entities, generally promote them. Stock exchange, production lines, standard agencies and general netted systems are meaningful examples of such management process.

Generally additional constraints are introduced (such as hierarchies, rules, operating modes, flow constraints, corporate charts, standards, modular design, network protocols, procedures embedded in information systems) in order to guarantee a required behavior (at least to render such behavior highly probable) by partially renouncing some goals related to efficiency, control and flexibility. Group technology is an example of this pattern. *Group technology* was introduced in the 1960s, mainly by Burbidge, as a product type organization for manufacturing, in which a factory is progressively divided into departments or groups which complete all the items they make without back-flow or cross-flow between them. In a group technology approach, there is a total division of machines and other processing facilities into groups and of all the parts made into associated families. Devices are grouped by technology type and may operate so that the management of each group can be organized efficiently in a homogeneous environment. This architecture often produces intricate routing patterns of parts and components.

Additional constraints are needed to manage complex systems, but they introduce rigidities. In order to overcome this problem, modern firms have begun to reorganize and decentralize their large organizations, introducing procedures configuring new pathways that allow project-centered groups to form rapidly and reconfigure as circumstances demand. This flexibility is also referred to as *organizational fluidity*: "Fluid organizations display higher levels of cooperation than groups with a fixed social structure. Faced with a social dilemma, fluid organizations show a huge variety of complex cooperative behaviors caused by the nonlinear interplay between individual strategies and structural changes" (Mainzer 1994).

The history of organized systems is largely determined by the way the different systems have modified the additional organization constraints originally introduced to deal with complexity in order to obtain greater efficiency. The change is driven mainly by market challenges and technical innovation, in part by internal reasons, but also by the cultural influence of the scientific and technical environment. Organizational constraints take on very different forms in different environments. Burbidge (1996) proposes a classification of industries based on the ratio of the number of material types (input) to the number of finished-product types (output). In process type industries, such as oil refineries or power plants, those numbers are both small; a small number of input types are transformed into a small number of output types. In the so-called implosive industries, mostly handicraft activities such as potteries or bakeries, the number of inputs is small, whereas the number of outputs is large; a small number of input types are transformed into many different products, sometimes in a single copy. In the so-called explosive industries, such as automobile or electronics, the number of inputs is large, whereas the number of outputs is small; this is the classical manufacturing plant, where many parts and components are transformed into a relatively limited number of products. In the so-called square industries, such as logistic or medical services, both the number of inputs and the number of outputs are large; this corresponds to open shops or other service centers. Note that the terms implosive and explosive are used in a typical industrial meaning: the size considered explosive is the length of the bill of materials corresponding to a given output, in an inverse design-oriented perspective, opposite to the production-oriented flow of materials.

There are further classifications for each of the previous classes. For example, in order to evaluate the different production control methods and to improve manufacturing efficiency, George De Albert Babcock (1875–1942) proposed a classification into seven major categories, based on the number of pieces to be produced: one or a small number never to be reproduced, repeated orders at irregular intervals for few or many pieces, repeated orders at uniform intervals for few or many pieces, continuous orders for the same piece (Babcock 1917).

There is a serious organizational dilemma in the design of high performance production units: to increase productivity tight coupling is recommended, but an efficient tight coupling management requires centralization; on the other hand, to be effective and flexible, interactive complexity is generally needed, but interactive complexity requires decentralization. Rail transport, airways, assembly lines, are all tight coupling frames built to simplify interactions. Universities, research and development groups, welfare agencies, are all loose coupling organizations with complex interactions. Aircraft, space missions, DNA, all have both tight coupling and complex interactions simultaneously.

The new simulation tools introduced in the late 1980s opened up the possibility of a more detailed analysis of the influence on the overall behavior of a decision process based on different organizational and procedural constraints, grounded on a large set of experiments made under different hypotheses and scenarios. This has produced a wide variety of organized structures and company procedures, that are however generally based on a few relatively simple principles. The best compromise between centralized and decentralized systems is probably the most important problem to have been thoroughly analyzed by the new techniques: global modeling, simulation and optimization, for the centralized systems; local optimization, autonomous agents or holons, for the decentralized ones.

Holon is a term introduced in 1949 by Arthur Koestler (1905–1983), and analyzed, from the point of view of hierarchies and stable intermediate forms in living organisms and social organization, in his well-known book *The ghost and the machine* (1967). Holon indicates a basic unit of biological and social entities with two fundamental features: it is a well defined entity and it is part of a larger system; a holon has a precise identity, can be easily separated from the rest of the system, its capability to operate depends on its interactions with the rest of the system, and it can often be split into more elementary units. A holon does not necessarily have thinking capabilities, and its goals and constraints depend, at least partially, on the rest of the system. This concept has been used to characterize a decentralized approach to production systems, called *holonic manufacturing systems*, which were popular in the early 1990s.

If a holon has the capability to think, to exchange information and to operate in the frame of an organized system, it is often called an autonomous agent. More precisely: *autonomous agents* are entities of an organized system, able to decide, to exchange information and to implement a sequence of actions, generally with a local knowledge of the situation and by suitable interactions with other entities.

As far as the production environment is concerned, the evolution of production systems has generally been characterized by a sequence of local adjustments, where portions of the whole system, in many cases only partially known, evolve, interacting with the neighboring subsystems, according to different criteria. The configuration of machines and infrastructures puts severe constraints on the overall evolution and in many cases determines the output. The basic elements of the system are often relatively simple and well structured; the whole system grows, driven by evolving internal requirements and client-server type exchanges with other elements, becoming more complex at each step.

In practice, it is a widely accepted fact that complex systems will evolve from simple systems much more rapidly if there are stable intermediate forms. The resulting complex systems will be hierarchic and their elements are subject to control by multiple higher authorities. The autonomous self-reliant units, which have a degree of independence and handle contingencies without asking higher authorities for instructions, are stable forms, which survive disturbances and provide the proper functionality to the system.

The resulting netted system generally forms complex patterns and is difficult to manage and control as a whole, although in practice some decision rules, typically developed along with the organization, allow people to act in many cases with a good coordination with the rest of the company. In practice, individuals are embedded into a suitable environment: individual memory contains some basic rules to react to inputs and the environment provides the right inputs to drive the individual behavior towards a high-performance operating mode. The environment in which the individuals operate has evolved in time by eliminating low-profile and promoting high-profile inputs; the resulting information pattern is sometimes referred to as *organization memory*.

More generally, in order to cope with the large number of elements and interactions, two general widely used criteria are *decomposition* and *standardization*: large subsystems are divided into smaller parts (as far as possible weakly interconnected) easier to manage and a set of standards on activities, network structure, communication protocols and information flows, are defined to manage the interactions. In particular, decomposition is a basic tool for actions aimed to modify the system. In fact, the effort needed to change a system is typically proportional to the size of the organization times the speed of the change (we can indicate such a quantity as *organization moment*); if we require significant changes in a short time, the quantity of effort we must be able to produce (and the corresponding quantity of money and people) is high; if we have a limited budget, only small changes can be made at each step.

Parts interchangeability expanding around 1800, *scientific management* introduced around 1900, and the diffusion of *total quality concepts* in the second half of the 20th century are by and large based on a structured and suitable application of those criteria (and on the development of the corresponding technology, such as mills, productivity measures and information systems, respectively). In particular, the evolution of production control in the last century depended on the availability of reliable, timely and high quality information from the shop floor. The borderline can be set around 1980 (McKay and Buzacott 1999): before this date the common tracking technique was manual, after this date, bar coding became widespread and PC-based systems became feasible. Up to 1980 traditional production control methods (and organizational structures) were substantially adequate to the problems posed by large, relatively stable, mature industries, which dominated the manufacturing markets. The management of complexity was basically achieved by making minor improvements in well-established patterns, which had proved effective in previous experience. Instability, uncertainty and network-type effects became significant aspects of industry management in the last part of the century.

Some design principles, such as *feedback, learning* and *case-based reasoning*, widespread in the last decade of the 20th century and today in the organization environment to cope with the difficulty of controlling real world phenomena, have their origin in a strictly technical environment. Feedback found its first industrial application in the second half of the 18th century to control stream power to supply usable energy to the new manufacturing industry. Learning and case-based reasoning were approaches used from the very beginning of the Industrial Revolution: the idea was to use other people's achievements in order to improve the way of dealing with specific transformation processes or more general problems. *Benchmarking* and *reengineering* are two well-known modern examples of this approach. One of the first industrial examples of case-based reasoning is the design of the first production line at a Ford plant in Highland Park, near Detroit, around 1913, where the benchmarking idea came from a overhead trolley that the Chicago packers used in dressing beef (Ford 1922).

A major aspect characterizing the different design criteria and principles is the level of knowledge required concerning the system to be dealt with. Feedback-based, evolutionary, learning, and case-based strategies require detailed external information on the input and output, but only general aggregate information on the internal behavior. Simulation and optimization strategies require detailed internal information on all the elements of the system. Autonomous agents are a compromise between the two approaches: detailed information is required on the network (its structure, the interaction protocols and all inputs and outputs of the agents), but only general and aggregate information on the internal behavior of the individual agents; moreover, new agents can be typically added and/or dropped without the need for extensive reorganization.

9.4 Decentralized Management of Complex Organizations

The decentralized management of production systems is a general trend of the last half century, supported by appropriate technologies (such as local numerical control of machines, powerful information and communication systems) and driven by the growing demand for flexibility and reduced lead-time. Modern production networks try to meet those requirements and accept the reduced power of control by top management, which is generally the price to pay. In a network environment, the way the top management controls the behavior of the company is by buying and selling activities, only seldom by direct actions aimed to modify the way a single entity operates. In fact, it is sometimes easier and more effective to sell a branch of a company, than to modify its behavior; on the other hand, it could be faster and cheaper to hire an established group of expert people, than to build-up such a group from the inside.

Network flow analysis is a powerful tool in the evaluation and design of production systems and more generally of netted companies. This fundamental tool of the mathematical theory of programming and allocation was developed from the early 1950s on and has produced a wide range of applications in organized systems. Typically enough, it was motivated by the "very practical problem of transporting a commodity from certain points of supply to other points of demand in a way to minimize shipping costs" (Ford and Fulkerson 1962); but the mathematical models provided by network flow analysis are based on graph theory, a research field in discrete mathematics whose origins go back to the 18th century. Network flow analysis was independently developed in various fields in order to manage different kinds of complex systems. Nevertheless, some unifying basic concepts exist in more or less all the applications, which have mutually influenced each other.

An important idea common to most network flow applications is to evaluate the behavior of a system formed by several parts or subsystems, by guaranteeing for each part the matching of a set of compatibility constraints determined by the behavior of other parts. This interdependence produces a set of complex patterns, which can in some cases be controlled by network flow techniques. In particular, there are several frames to evaluate and control the flow of parts, when the interactions are basically only movements of physical or informational items among sections of the system, as in flow assembly lines, supply chains or general client-server relations.

The networks studied in literature are typically linear, mostly single-commodity and single decision-maker (and, in practice, some of the simplest properties of network flows have sometimes proved useful for the first stage design and management of complex systems). On the other hand, the networks related to complex organization systems are typically non-linear, multi-commodity, with indivisibility requirements and distributed decision-making. Moreover, they are often only partially defined, with ill-defined boundaries, incompletely known relations with external entities and human based portions of the system, difficult to represent in quantitative terms. The overall system is a multi-layered network, where the nodes represent autonomous subsystems or individual entities (often referred to as autonomous agents or general holons). The arcs represent different kinds of relationship among agents; each node corresponding to a subsystem has a different level of autonomy and can generally be represented as a network.

The original *centrally controlled networks*, where all actions are decided centrally and the agent acts locally without any real decision-making power and, in particular, without the capability to modify its behavior owing to interactions with the environment (such as most traditional chemical plants or some rigid manufacturing/assembly lines), evolve towards higher levels of autonomy. The emerging agent autonomy ranges from the *autonomy of action*, where the agent has the capability to adapt its actions to the response of a variable environment or to other kinds of inputs, on the basis of a given set of decision rules (as in standard Flexible Manufacturing Systems or in Automated Guided Vehicles); to the *autonomy of decision*, where the agent has the capability to select the best decision (and to perform the related actions), on the basis of a given set of objectives and constraints (as in organization units responsible for the maintenance and operability of a set of devices, for a group of end-users); to the *autonomy of objective*, where the agent has the capability to choose its own objectives, on the basis of a given set of general external constraints and operating modes (as in many design groups and research and development structures); to the *autonomy of business*, where the agent acts as an independent firm, negotiates solutions with other entities, promotes joint ventures and partnerships.

The typical goal most of the companies pursue in the production systems environment, in order to optimize the production process and maximize performance, is autonomy of decision, but there are several significant exceptions. The *process networks* realized in many manufacturing and assembly lines are generally based on an autonomy of action level; numerically controlled machines and industrial robots have the capability of a large set of operations that they can choose from according to the state of the field and the inputs from the production manager, but they cannot modify the rules. The agents forming a *supply chain* in some cases have autonomy of objective; they can, at least partially, modify their objectives to suit the situation in the field. Many *business units* of large companies or a *netted virtual factory* formed by several interconnected small business units have autonomy of business; they are, from many points of view, really independent firms.

The design of such networks is an on-going process, where on one hand a suitably large set of agents performing basic functions at different levels of autonomy is built up by different field specialists, and on the other different partial multi-layered patchworks utilizing the available basic elements are first tried out (often using simulation tools) and then implemented, in order to obtain high performances, by several (often independent) systems and organization designers: "Organizational design addresses who does what, what part of the business an individual employee has interaction with, how authority is distributed, how information will flow through the company, and how personnel will move through the company" (McKay and Buzacott 1999). Each autonomous agent and each designer has their own domain of decision and action, their own objectives (either given or built up locally) and constraints. Moreover each autonomous agent is characterized by a kernel of essential technologies, where it is highly competitive, and by a set of market drivers to be exposed to other concerned entities.

The objectives are generally layered into several levels, depending on the time span considered and the autonomy level. The lower level objectives typically include the system performances in nominal operating conditions (i.e. assuming a given stable environment and no disturbances); production rate, system capacity, unit cost, productivity, effectiveness, defect-rate, lead-time and efficiency are the corresponding standard performance indexes. The upper level objectives typically include the system behavior under variable operating conditions and different scenarios; different types of flexibility, robustness, total cost, operability, controllability and other total quality indexes are among the corresponding most widely used performance indexes.

The decentralized management of an organization has always been a major issue for the top managers of large private and public organizations (manufacturing and service companies on one hand, military, financial and public service organizations on the other), but only in the last twenty years of the 20th century was this issue formalized and quantified in a relatively complete frame.

The organization chart with the multi-division organization proposed by Sloan in 1921 for General Motors was basically concerned with product choice and financial flows management: "the basic elements of financial control in General Motors are cost, price, volume, and rate of return on investment" (Sloan 1963). The logistic support organization during the Second World War was designed to guarantee local operability under harsh wartime conditions. The distributed fail-soft system configurations of the 1960s and 1970s were mainly concerned with reliability and availability: air traffic control systems, telephone networks, and some flexible manufacturing plants are well known examples of reliable tight-coupled fail-soft systems. The various *Management By Objective* organizations, widespread in the 1970s and 1980s, were concerned with the problem of how to attribute the local goals to low and middle management in order to achieve high global performance. Many *total quality* organizations of the 1980s were mainly concerned with the ways the end-user requirements could be taken into account and affect the supply chain.

Modern production networks supported by sophisticated information systems and autonomous agent based organizations achieve a certain degree of compatibility and allow decentralized decision processes with high local effectiveness, low lead time, a certain degree of flexibility and global feasibility. As far as efficiency is concerned, a fairly efficient use of resources can usually be obtained in the short term only by careful centralized planning, generally supported by centrally managed resource administrators and network supervisors. An "autonomous agents network" leads to a flexible system, with self-adjustment on defined problems and data-based operator-centered process improvement for many undefined problems. The high-speed high-quality production process, monitored by time-relevant performance metrics, is characterized by an increase in frontline responsibility, with direct data support, thereby reducing the number of transactions.

Network organizations take on different forms in different environments. In craft-type industries, work is organized around specific projects and involves the temporary cooperation of varying combinations of skilled workers and other types of production resources, such as in large construction projects or in film production. In small firms homogeneous clusters, such as the low-tech textile companies of the northern Italian industrial districts or the high-tech semiconductor firms in Silicon Valley, the production system is geographically located and different types of horizontal and vertical interactions take place.

9.5 Open Systems

Some important aspects of complexity can be analyzed only by considering the system embedded in a larger environment, with many entities and intricate mutual influences. This approach could increase the overall complexity of the system, but could also simplify some control actions directed towards the system. In fact, it is generally difficult to model the relationship between the entities of the organized system and what is considered environment. On the other hand, a management of complexity pursued by introducing boundaries, using external constraints, implementing external-driven actions, can in many cases be realized without a complete knowledge of the system, on the basis of past experience and standard reactions.

In the conceptualization proposed by Henry Mintzberg to represent general organization systems there are six basic elements (Mintzberg 1983): three of them (top management, in charge of the strategic decisions, intermediate management, dedicated to resource administration, and low management, in charge of the operations) form the basic structure of the hierarchy, which transforms inputs (resources) into outputs (products); one is the environment, which contributes to defining the mission, the feasibility of actions and the global performance of the system; another is the support structure, responsible for financial and personnel administration and the management of the basic logistic infrastructures; and the last one is the techno-structure, formed by engineering departments, research and development department, units responsible for technical standards and strategic marketing.

The first three elements, from an organizational point of view, are mainly internal; they have connections with external entities such as suppliers and clients, on a client-server basis. The last two elements have strong structural links with the environment: the former with financial and general services suppliers, trade unions and local administrations; the latter with other companies, technical services, parts and devices suppliers, scientific and technical institutions. External links play a fundamental role in determining the behavior of portions of the system; they reduce the degree of freedom of internal entities, but increase stability and quality.

Simon defined a complex system as "one made up of a large number of parts that have many interactions" in his paper on "The architecture of complexity" (1962), while James D. Thompson, in his textbook *Organizations in action* (1967), defined a complex organization as "a set of interdependent parts, which together make up a whole that is interdependent with some larger environment". The concept of *open system* was taking its first steps in organization theory: an entity was said to be open when it exchanged resources with the environment and system when it consisted of interconnected components working together.

Several concepts, originating mainly from physics and biology, are used to analyze open systems. The concept of entropy, introduced in physics to explain some basic relationships of gas dynamics, and the concept of biological evolution have been widely utilized in other fields and, in particular in economy and in organization.

Only in the late 1980s did the modeling of open systems become a well-established research field in organization theory. To support the analysis, a new field was developed: *complex adaptive systems* (cas). Instead of explaining a given behavior by traditional input-output analysis, cas "asks how changes in the agents' decision rules, the interconnections among agents, or the fitness functions that agents employ produce different aggregate outcomes" (Anderson 1999). Adaptive denotes change, but how can the different kind of changes be classified? Cas can modify their structure to adapt to external inputs, changes in other entities or variable internal requirements. The changes can be improvements as measured against some standard, or simply changes in response to other changes (of other connected entities or the environment, whatever this means). If the changes are improvements, the performance can be evaluated on the basis of external standards (e.g. decided by an external authority) or of internal standards (e.g. output from an internal decision process, embedded in the system).

Following the Murray Gell-Mann approach to the problem: "cas encode their environment into many *schemata* that compete against one another internally" and "complexity is the length of the schema needed to describe and predict the properties of an incoming data stream by identifying its regularities" (Gell-Mann 1994). In this framework, encoding a system into one or more formal schema, compressing a longer description into a shorter one that is easier to grasp, makes the management of complexity possible. Each schema is an approximation of reality and explains some aspects of the system. Different schemata typically explain different aspects and are selected on the basis of performance in the field and of the objectives of the management process.

The basic idea of schemata is that the evolution of a complex system is not a continuous process; a new way to deal with the current problems (and the corresponding organization structure) replaces the old one completely, in a short space of time, with an organizational break. Using the words of Thomas Kuhn: "[...] scientific revolutions are [...] non-cumulative developmental episodes in which an older paradigm is replaced in whole or in part by an incompatible new one. [...] In both political and scientific development the sense of malfunction that can lead to crisis is a prerequisite to revolution. [...] As in manufacturing so in science – retooling is an extravagance to be reserved for the occasions that demand it. The significance of crises is the indication they provide that an occasion for retooling has arrived" (Kuhn 1962).

The fact that a system is open has a strong influence on the ways we can deal with complexity and on the possibility of managing complexity. Ilya Prigogine claims, "when a system is open to receiving energy [and information] from the outside, it will tend to create order. When a system becomes closed it will decay into maximum disorder and chaos. [...] At any level of analysis, order is an emergent property of individual interactions at a lower level of aggregation" (Prigogine 1984).

The approach based on schemata offers new ways to produce high performance evolution patterns and is a powerful tool for the analysis of decentralized systems based on autonomous agents. *Technological convergence* is a typical phenomenon adequately explained by the schemata approach. In each company, technology innovation tends to move in the direction of the winning technologies developed by the most successful competitors. The corresponding winning schemata replace the old ones. The competition/substitution process is often a creative process: competition between alternative schemata can modify schemata or produce completely new schemata (and new organization structures). A *recombination* process, similar to biological evolution, generates new schemata, where positive elements of both the competing schemata are combined to improve the output. In many cases, a process merging the original agents, which are combined to create a single organizational unit, can obtain this more effectively than by trying to modify the schemata used by the two agents. Some joint ventures and corporate mergers can be interpreted as a recombination process. The creation of task forces, interdepartmental groups and interdisciplinary teams can often also be interpreted in this way.

Netted *companies*, both in manufacturing and in service systems, are typical examples of open systems. They operate in two basic environments: large decentralized companies with an embedded network structure centrally coordinated with relevant global objectives (*central networks*) and large networks of small independent objective oriented companies coordinated by market type interactions (*distributed networks*). In both cases information technology and project organization play a fundamental role. The interactions among the *organization units* and with the outside organizations, in particular for central networks, are based on procedures, protocols and standards carefully defined by ad hoc organizations to support the so-called organization network. An *organization network* can be defined as a system of two or more organizations that exchange resources and negotiate decisions to reach objectives that cannot be reached by single organizations separately.

There are several significant differences between central and distributed networks. Organization units inside large companies have well-defined boundaries and rigorous protection of know-how, whereas networks of small companies have ill-defined and variable boundaries; moreover new ideas tend to become available to all members of the network in a short space of time. The added value of a distributed network tends to be increasingly related to the capability to transform ideas into a business and to maintain good personal relationships, rather than to new ideas; for a successful netted company, technology innovation cannot be separated from innovation in organization and in market relationships. As far as personal incentives are concerned, large companies tend towards broadband long-range personal incentives, whereas small independent units tend towards narrowband project-oriented personal incentives.

In traditional organizations, information on the state of the system flows bottom-up and decisions flow top-down. An organization network is generally characterized by a lean organization with a basic information symmetry (information and decisions flow uniformly in the system), which can be supported by different types of organization structures, although there are some basic features that are typically present to make a network successful. Central networks often have a process type of organization and highly structured and standardized supply chains. The information flow is very rich, highly structured and involves a limited number of known entities, people and groups. Distributed networks are characterized by a continuous reassembling of the supply chain with a flexible on-line management of complexity based on direct negotiation. The information flow is relatively poor, low structured and involves a large number of often unknown entities, people and groups. Some frequent general features are: decentralized management of operations and decisions with a high capability for integrating different local actions, effective on-line control over strategic resources, high flexibility and robustness, adaptive project-oriented organization structure, weak institutional links, short and limited power lines, high efficiency in core operations and capability to create new markets, low cost of change, low entry barriers, low negotiating power, low lead time.

A network is more suitable than a centralized system for coping with an evolving environment. Changes could be limited to flow management aspects, such as rerouting and rescheduling (leading people to deal with process complexity problems), or could confront intermediate level aspects, such as topological change or downsizing, or could be concerned with general system aspects, such as upgrading or embedding of the network in a larger environment. Due to the rapid evolution of the network, the limited information, the uncertainty about goals and constraints, significant problems stand in the way of obtaining a reliable global forecasting of the state of the network and of guaranteeing global stability. It could prove difficult to optimize operations and to make a rational choice of the best solutions to adopt. The optimization process is increasingly concerned with system aspects, management functions and organization structure. Instead of the choice of solutions, the most important aspect becomes the choice of operating mode, i.e. the choice of the most influential constraints.

A significant example is *just in time*, first introduced by Toyota, which then became a key element of almost all industrial strategies of the second half of the 20th century. It is a method of production flow control, in which products are only assembled when they can be delivered to customers, parts are only made when they are needed for assembly or for delivery, and the receipt of purchases is only accepted when they are immediately required for further processing. Just in time modifies production constraints and goals, introduces new processing functions, new ways of operating and requires new types of skills.

In order to deal effectively with system complexity problems, the network designers bestow a privilege upon modularity, symmetry and regularity. On the other hand, those concepts lead to more vulnerable networks: a problem related to a particular pattern spread over the entire network. In order to reduce the potential damage, *differentiation* (sometime referred to as *biodiversity*) could be much more effective than many barrages and net defenses.

Acknowledgments

This paper is basically a collection of ideas of some prominent scientists of the 20th century; the way to collect these ideas and the few new ones are the result of many fruitful discussions with Fernando Nicolò (often during beautiful mountain trekking). Ana Millán Gasca, who urged me to write the paper, has carefully revised several preliminary versions, suggesting many significant improvements.

References

Anderson Ph., 1999, "Complexity theory and organization science", *Organization Science*, 10: 216–232.

Babcock, G. D., 1917, *The Taylor System in Franklin Management: Application Results*, New York, The Engineering Magazine.

Burbidge J. L., 1996, *Period Batch Control*, Oxford, Clarendon Press.

Burks, A. W., Goldstein, H. H., and von Neumann, J. L., 1946, *Preliminary discussion of the logical design of an electronic computing instrument*, Report prepared for U.S. Army Ordnance Department under contract W-36-034-ORD-7481, in: *Papers of John von Neumann on Computing and Computer Science* (Aspray, W. and Bruks, A., eds.), Cambridge (Mass.), MIT Press, 1987: 97–142.

Clippinger, J., 1999, *The Biology of Business: Decoding the Natural Laws of Enterprise*, San Francisco, Jossey-Bass.

Coes, H. V., 1928, "Mechanical scheduling", in: *110 Tested Plans that Increased Factory Profits; Ideas Selected from the Pages of Factory and Industrial Management, as of Particular Value in Practical Factory Management* (Dutton, H. P., ed.), Chicago-New York, McGraw-Shaw.

Cook, S. A., 1971, "The complexity of theorem proving procedures", in: *Conference Record of the Third Annual ACM Symposium on Theory of Computing, Shaker Heights, Ohio*, 3–5: 151–158.

Cowan, G. A., Pines, D., and Meltzer, D. (eds.), 1994, *Complexity: Metaphors, Models and Reality*, Reading (Mass.), Addison Wesley.

Deming W. E., 1993, *The New Economics for Industry, Government, Education*, Cambridge (Mass.), Massachusetts Institute of Technology, Center for Advanced Engineering Study.

Deming W. E., 1950, *Elementary Principles of the Statistical Control of Quality*, Tokyo, Nippon Kagaku Gijutsu Renmei.

Deming W. E., 1986, *Out of the Crisis*, Cambridge (Mass.), MIT Press.

Dorfman R., Samuelson, P., Solow, R., 1958, *Linear Programming and Economic Analysis*, New York, The Rand Corporation/McGraw Hill.

Fayol, H., 1918, *Administration industrielle et générale*, Paris, Dunod (English translation, *General and Industrial Management*, London, Pitman, 1949).

Ford, H., 1922, *My Life and Work*, New York, Doubleday.

Ford L. R. and Fulkerson, D. R., 1962, *Flows in Networks*, Princeton (N. J.), Princeton University Press.

Gantt, H. L., 1910, *Work, Wages and Profits*, New York, The Engineering Magazine.

Gantt, H. L., 1916, *Industrial Leadership*, New Haven, Yale University Press.

Gantt H. L., 1919, *Organizing for Work*, New York, Harcourt, Brace and Howe.

Gantt, H., 1961, *Gantt on Management; Guidelines for today's Executive*, New York, American Management Association.

Garey, M. R. and Johnson, D.S., 1979, *Computers and Intractability: Guide to the Theory of NP-completeness*, San Francisco, Freeman.

Gell-Mann, M., 1994, "Complex adaptive systems", in: Cowan, Pines, and Meltzer 1994.

Ishikawa, K., 1976, *Guide to Quality Control*, Tokyo, Asian Productivity Organization.

Ishikawa, K., 1990, *Introduction to Quality Control*, Tokyo, 3A Corporation,.

Karp, R. M., 1972, "Reducibility among combinatorial problems", in: *Complexity of Computer Computations* (Miller, R.E. and Thatcher, J.W., eds.), New York, Plenum Press.

Kerzner, H., 1992, *Project Management*, New York, Van Nostrand Reinhold.

Koestler, A., 1967, *The Ghost in the Machine*, London, Hutchinson.

Koopmans, T. C. (ed.), 1951, *Activity Analysis of Production and Allocation*, New York, Yale University Press/Wiley.

Kuhn, Th., 1962, *The Structure of Scientific Revolutions*, Chicago, University of Chicago Press.

Lenstra, J. K., Rinnooy Kan, A. H. G., and Schrijver, A. (eds), 1991, *History of Mathematical Programming. A Collection of Personal Reminiscences*, Amsterdam, CWI/North Holland.

Leontief, W., 1966, *Input-output Economics*, New York, Oxford University Press.

Mainzer, K., 1994, *Thinking in Complexity: The Complex Dynamics of Matter, Mind, and Mankind*, Berlin-New York, Springer.

Mayo, E., 1945, *The Social Problems of an Industrial Civilization*, Cambridge (Mass.), Harvard University Press.

McKay, K. N. and Buzacott, J. A., 1999, "Adaptive production control in modern industries", in: *Modeling Manufacturing Systems: From Aggregate Planning to Real-time Control*, (Brandimarte, P. and Villa, A., eds.), Berlin-NewYork, Springer.

Mintzberg, H., 1983, *Structure in Fives. Designing Effective Organizations*, Englewood Cliffs, N. J., Prentice Hall.

von Neumann J., and O. Morgenstern, 1944, *Theory of Games and Economic Behavior*, Princeton (N. J.), Princeton University Press.

Ohno, T., 1988, *The Toyota Production System: Beyond Large-scale Production*, Cambridge (Mass.), Productivity Press.

Perrow, Ch., 1972, *Complex Organizations: A Critical Essay*, Glenview, Ill., Scott, Foresman.

Prigogine, I. and Stengers, I., 1984, *Order out of Chaos: Man's New Dialog with Nature*, New York, Bantam Books.

Pugh, D. S., 1971, *Organization Theory*, London, Penguin Books.

Riggs, J. L., 1970, *Production Systems: Planning, Analysis, and Control*, New York, Wiley.

Scott, W. R., 1981, *Organizations*, Upper Saddle River (N. J.), Prentice Hall.

Simon, H., 1962, "The architecture of complexity", *Proceedings of the American Philosophical Society*, 106(6).

Simon, H., 1969, *The Sciences of Artificial*, Cambridge (Mass.), MIT Press.

Simon, H., 1982, *Models for Bounded Rationality*, Cambridge (Mass.), MIT Press.

Sloan, A. P., 1963, *My Years with General Motors*, New York, Doubleday.

Taguchi, G., and Konishi, S., 1987, *Orthogonal Arrays and Linear Graphs: Tools for Quality Engineering*, Dearborn (Mich.), American Supplier Institute.

Taylor, F. W., 1911, *The Principles of Scientific Management*, New York, Harper and Brothers.

Thompson, J. D., 1967, *Organizations in Action*, New York, McGraw Hill.

Index of Names

Authors

Stuart Bennett
Department of Automatic Control
& Systems Engineering
University of Sheffield
Mappin Street
GB-Sheffield S1 3JD

Amy Dahan Dalmedico
Centre Alexandre Koyré
Pavillon Chevreul
Museum d'Histoire Naturelle
57 rue Cuvier
F-75005 Paris

Giorgio Israel
Dipartimento di Matematica
Università di Roma "La Sapienza"
Piazzale A. Moro 2
I-00185 Roma

Eberhard Knobloch
Institut für Philosphie,
Wissenschaftstheorie, Wissenschafts- und
Technikgeschichte
Technische Universität Berlin
Ernst-Reuter-Platz 7
D-10587 Berlin

Antonio Lepschy
Dipartimento di Elettronica e Informatica
Università degli studi di Padova
Via Gradenigo 6/a
I-35131 Padova

Ana Millán Gasca
Dipartimento di Matematica
Università di Roma "La Sapienza"
Piazzale Aldo Moro 2
I-00185 Roma

Evgenii F. Mishchenko
Steklov Mathematical Institute
Russian Academy of Sciences
Gubkin St. 8
RUS-Moscow 117966, GSP-1

Alexandr S. Mishchenko
Steklov Mathematical Institute
Russian Academy of Sciences
Gubkin St. 8
RUS-Moscow 117966, GSP-1

Fernando Nicolò
Dipartimento di Informatica
e Automazione
Università di Roma 3
Via della Vasca Navale 79
I-00146 Roma

Dominique Pestre
Centre Alexandre Koyré
Pavillon Chevreul
Museum d'Histoire Naturelle
57 rue Cuvier
F-75005 Paris

Umberto Viaro
Dipartimento di Energia Elettrica,
Gestionale e Meccanica
Università di Udine
Via delle Scienze 208
I-33100 Udine

Andrzej Wierzbicki
National Institute of Telecommunications
Szachowa 1
PL-04-894 Warsaw

Mikhail I. Zelikin
Steklov Mathematical Institute
Russian Academy of Sciences
Gubkin St. 8
RUS-Moscow 117966, GSP-1